EASY ESSENTIAL OIL CHEMISTRY

Unlock the Healing Potential of Essential Oils

JIMM HARRISON

Easy Essential Oil Chemistry

Cover and interior design by Jane Ashley

To contact the publisher, visit
FLOWEROFLIFEPRESS.COM

To contact the author, visit
JIMMHARRISON.COM

Library of Congress Control Number: 2017954247
Flower of Life Press, Old Saybrook, CT.

ISBN-13: 978-0-9863539-3-2

Printed in the United States of America

Praise

"This book is an excellent dive into a deeper understanding of Essential Oil chemistry. The explanations and diagrams in this book really help to clearly and concisely explain complex chemistry in a very enlightening and practical way. This book will help the user have a better understanding of the chemistry and therapeutic potential of essential oils and provides a nice framework in which to apply this information for practical applications."

–Sheila Kingsbury, ND, RH (AHG),
Chair of Bastyr University Botanical Medicine department

"Jimm Harrisons introduction to the chemistry of essential oils is a most valuable tool for the aromatherapy afficionado to design therapeutic formulations of essential oils. This holds true despite current critical discussions of the validity of Functional Group Therapy. It is true that the functional group concept does not explain the intricacies of biological interactions of essential oil components with target proteins in the cell. However, the French-style approach to essential oil chemistry has been validated by successful applications of thousands of users. This French-style approach is most useful when it comes to essential oils which are rich in Monoterpene components.

For instance Ravintsara, Eucalyptus radiata and Niaouli—all rich in Monoterpene alcohols, Monoterpene hydrocarbons and Cineole—are commonly combined to counteract upper respiratory viral conditions. Again, while chemistry alone does not make statements about biological interactions it is indicative of physico-chemical parameters (of the respective components), which as a result of polarity and lipophilic nature alone lead to different interactions with targets in the cell membrane."

–Dr. Kurt Schnaubelt, *Pacific Institute of Aromatherapy*

Dedication

I thank all the students who, through their questions and differences in learning styles, have helped me to develop the lessons in this book.

For Kim–Thanks for the love, support, inspiration, craziness and desires that move us forward in life.

And thank you to the following:

Kim Buckner, *Research and Associate*; PJ Hanks, *Editorial Associate*; Barbara Beach-Moody, *Content Assistant.*

Contents

Introduction

If you're reading this, you must have a passionate interest in learning more about the incredible healing potential of essential oils—and how to best capture that power. To really know essential oils, it is critical you learn about their chemical compounds. The chemistry is where the healing resides, how the oils are studied in scientific evaluation, and how they determine the quality of the final product. The chemistry is an important part of essential oil understanding.

This book is especially for those who struggle with the essential oil chemistry that is often taught, talked about, and written along with essential oil descriptions. What are you able to do with this chemistry information as presented? Nothing. The relevancy connecting the knowledge to the use is missing. If you are a chemist, you may be able to turn a compound name, like 3-Methyl-2-buten-1-yl acetate, into a compound drawing or model. Impressive. I don't have that skill. What will the chemist be able to do with that drawing to assist in more effective use of essential oils? Nothing. You, the chemist, are in the same boat with the rest of us.

You don't need to be a chemist or know chemistry to use essential oils. The chemical compounds are just words we use. These chemical words then get identified for a therapeutic use. There is a big difference between drawing compounds or naming every component of an essential oil and actually knowing the therapeutic activity and potential of those chemical compounds. This isn't about chemistry for the sake of knowing chemistry. It's about creating a system to make use of words, in this case chemical words. That is the motivation of this book: Teaching you how to take these words and place them into an easy-to-learn system to make your work with essential oils more accurate and effective.

I was taught about the chemical compounds and the importance of recognizing them, as best as can be done. I was introduced to the chemistry and an adaption of the structure diagram in the first few classes I took with Michael Scholes. Through a series of classes, books and teachers I was able to build a familiarity with some of the chemistry and the chemical families. I was deeply challenged to make use of the chemistry and wanted to find a way to establish the relationship between knowing the chemistry and using it therapeutically.

Dr. Kurt Schnaubelt is the inspiration for what I have developed in this book. Dr. Schnaubelt introduced me to the structure diagram as a system I could use to select essential oils. The structure diagram was originally presented by Pierre Franchomme in the book *L'aromathérapie exactement.* The Structure-Effect Diagrams in this book are based on diagrams by both Schnaubelt and Franchomme.

What follows is an evolution of development for teaching essential oils and the importance of learning the chemistry. I have taught this to a diverse group of students, some bio-chemists or Naturopathic doctors and some with zero chemistry or medical background. The diagrams and the teaching method here are focused on turning words into a useful therapeutic tool exclusive to essentials oils. It works for everyone.

~Jimm Harrison

Electron Microphotograph of Clary Sage

PHOTO CREDIT © POWER & SYRED

PART 1

The Structure-Effect Diagram: A Visual Model Showing the Chemistry and Composition of Essential Oils

Essential oils are a symphony of fragrant and healing chemical compounds. Their chemical structure tells a story of the plant and the oil's therapeutic potential, its scent and unique characteristics. Scientific analysis has dissected and evaluated the complex chemical structure of essential oils in a quest to better understand their healing potential. The Structure-Effect Diagram is a visual representation of the complex essential oil chemistry and a beneficial tool used to evaluate chemical properties and therapeutic application of essential oils.

Essential Oils Are the Plant's Protective System

Plants produce chemical compounds, called secondary metabolites, as protection from parasites (bacteria, virus and fungi), to harmonize with their environment (cold, heat, ultraviolet radiation) and to repair cellular damage. Secondary metabolites are also a communication system for plants, such as attracting insects for pollination or warning neighboring plants of parasitic invasion. Essential oils are secondary metabolites—as are other food and herbal compounds, like carotenoids, flavonoids and alkaloids—that are well known for health and wellness.

Clues to the potential therapeutic properties of oils comes from observing the originating plant. For example, a desert plant such as frankincense requires protection from the sun's harsh ultraviolet radiation (UVR). The plant will produce secondary metabolites to harmonize with the environment, developing compounds that will neutralize the damaging effects of UVR. Frankincense happens to be a good essential oil for protection from sun damage.

The Structure-Effect Diagram presents a picture of an essential oil's complex mixture of chemical compounds and assists in the selection of essential oils for known therapeutic properties. Knowing the properties attributed to the chemical families and the individual compounds creates a deeper relationship with essential oils. The diagram and knowledge of the properties associated with the chemical groups enables finely tuned essential oil selection for more confident treatment choices and application. It's the mixture of compounds in each essential oil that provides the overall healing personality to the oil.

Comprehension of the oil's chemical composition and synergistic compound mixture will help you identify multiple uses, contraindications and safety issues. This knowledge will also help navigate inconsistencies often written about regarding the use of essential oils. The presentation in this book is a guide, providing a visual representation of the chemical structure and potential healing within an essential oil using the Structure-Effect Diagram.

Chemistry is Used as a Guide to Therapeutic Use

The scope of essential oil therapy is much broader than the chemistry alone. Essential oil chemistry is a guide to the properties and therapeutic action of the oils and is used to make accurate therapeutic assessments. The Structure-Effect Diagram is the tool that helps to understand the compound groups and their properties.

Essential oils are biological substances containing a mixture that is much more than a series of independent molecules. The whole oil extends the therapeutic action beyond what is known about the individual molecules. This is an exciting synergy that holistically assists our body's healing needs. It's difficult to fully comprehend the synergistic effects. The properties of the compound families in each essential oil synergistically emerge to create a biological element to the therapeutic properties. This synergy cannot be defined by the individual chemistry alone. The structure diagram provides direction to the selection of essential oils and to anticipate synergistic activity.

Plants, and their production of secondary metabolites, were well established before the existence of humans on the planet. Human development evolved and is dependent upon eating plant-based foods—consuming secondary metabolites. Humans do not produce these compounds even though they support the life and health of the human being. Mammals evolved consuming secondary metabolites in the diet and using them as remedies to protect from parasites and environmental conditions. The human relationship with plant-based secondary metabolites, and the plants they come from, guides a logical use of essential oils for health and wellness.

Secondary Metabolites as Essential Oils: A Complex Mixture of Multi-Target Compounds

Essential oils are a complex mixture of multi-target secondary metabolites. Multi-target refers to the ability of the compounds to react with several cell receptors, unlike the one-dimensional, one-receptor action of human-made molecules often used in pharmaceuticals.

Molecules attach to cell receptors in a lock-and-key fashion. Neurotransmitters, hormones and other messenger molecules, or ligands such as serotonin and endorphins, have unique properties and shape, like a key, that will fit specific cell receptors, the lock. When the molecule binds, it sends a specific signal to the cell, triggering a behavior or chemical change. This is the action that drives, controls and directs the human body and all living organisms.

Essential oils are like a key that fits many locks (targets). This multi-target ability allows for very diverse action and therapeutic capability and may include unexpected healing results. The multi-target action also explains why bacteria do not easily become resistant to essential oils. The complex mixture of multi-target chemical compounds provide essential oils the potential to work in a synergistic and balancing way with the human body.

The Compound Groups

The majority of molecular compounds found in essential oils are classified, according to their atomic structure, into groups or families on the Structure-Effect Diagram. In essential oil chemistry it has been determined that each group has specific properties. The properties of each group coincides with the properties of the individual molecules in the essential oils. Much of the healing action of the essential oil is a result of the individual and combined synergistic properties of the molecular compounds it contains. Throughout this text the terms *group(ings)* and *family(ies)* will be used interchangeably.

The Structure-Effect compound groups are:

- Monoterpene (C10) Hydrocarbons
- Sesquiterpene (C15) Hydrocarbons
- Monoterpene Alcohols
- Sesquiterpene Alcohols
- Aldehydes
- Esters
- Ketones
- Oxides
- Phenols
- Phenylpropanoids (Acids, benzoic acid and cinnamic acid, are placed in the phenylpropanoid group. They have similar properties and are produced by the same biosynthetic pathway.)
- Ethers
- Lactones and Coumarins

What is the Structure-Effect Diagram?

The Structure-Effect Diagram is a visual representation of the chemical compounds and their potential therapeutic action. This diagram is used to create a visual representation of the compound structure of an essential oil, identifying its therapeutic potential. Molecular groups are represented on the Structure-Effect Diagram by oblong and circle shapes according to the range of their electromagnetic charge (above or below the horizontal line) and their polarity (polar, left or non-polar, right).

Pierre Franchomme and Daniel Pénoël introduced the diagram to the aromatherapy trade in their book *L'aromathérapie exactement.* The diagram is structured according to scientific analysis of the electromagnetic charge and polarity of the individual essential oil compounds.

Electromagnetic charge is the physical charge—positive, negative or neutral—produced by electrically charged objects, including molecules.

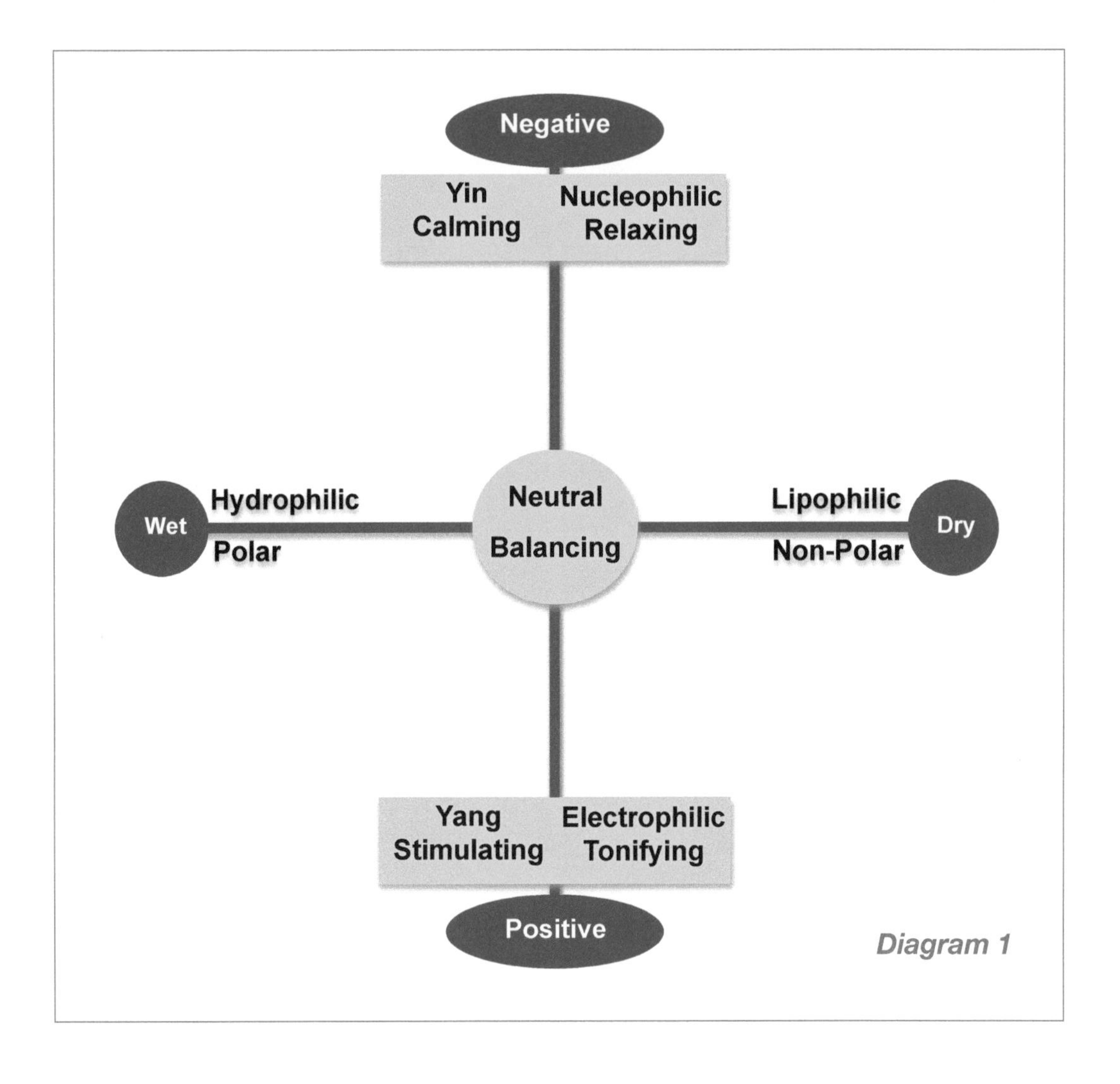

Diagram 1

The following paragraphs describe the conditions, or traits, that help to identify potential action and properties of the molecules in each group:

Electronegative and Calming

A molecular group, and the compounds contained in that group, positioned above the horizontal line have variation in electronegativity. The more negative the charge the more calming the compound.

Electropositive and Energizing

The more positive the electromagnetic charge of the group, the further to the bottom it is placed on the chart. Electropositive compounds are energizing or stimulating.

Polarity and the Ability to Bind to Water

Polar - The further to the left, or "wet" side, the group is placed, the more polar the group is (meaning it has more harmony and ability to bind with water).

Non-Polar - The further right, or to the "dry" side, the group is positioned on the diagram, the more non-polar the group is, meaning there is no bond with water at all, but it does have a strong bond with other fatty acids or oils (lipids).

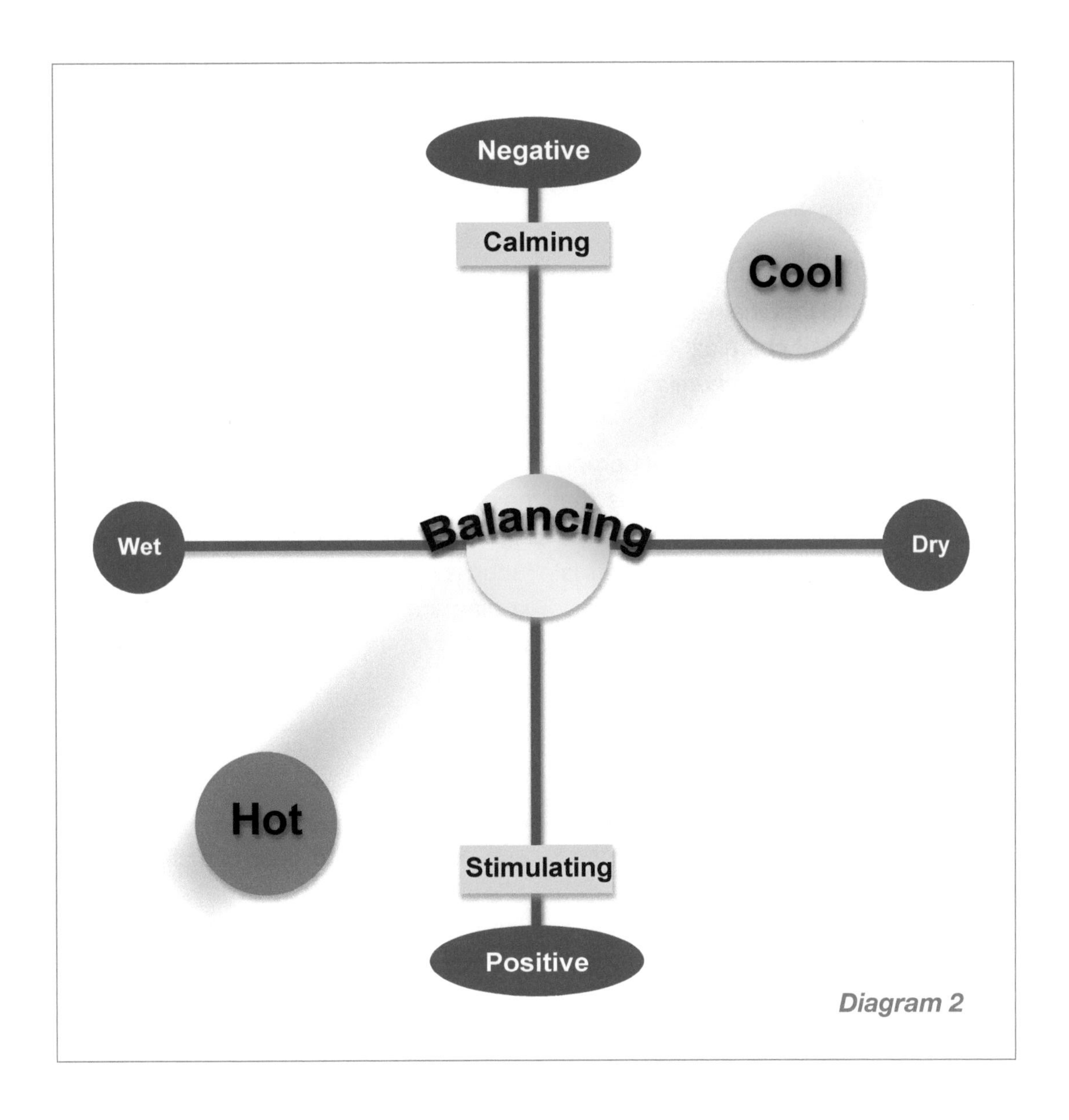
Negative
Calming
Cool
Wet
Balancing
Dry
Hot
Stimulating
Positive

Diagram 2

The Structure-Effect Diagram Layout

The Structure-Effect Diagram is used as a visual tool to understand the calming, stimulating, balancing, hot, or cooling quality of a group, and the essential oils that contain molecules from the group. Diagram 3 shows where the groups are positioned on the Structure-Effect Diagram.

Each Molecular Group Contains Many Compounds

Keep in mind that the essential oils contain a mixture of chemical compounds. Each individual chemical compound belongs to a chemical family, identified and classified by its chemical structure.

It is individual molecules that compose the essential oils. The Structure-Effect Diagram represents the groups that these molecules belong to. The visual shape of each group represents the range of where the individual molecules may appear on the diagram. The main focus is on the overall properties of the group, not the individual molecules. The diagram is a helpful visual guide for learning the structure of an essential oil relating to its compounds, the chemical family and therapeutic potential.

Positioning and Properties of the Molecular Groups on the Structure-Effect Diagram

The groups on the diagram are positioned according to their overall properties. The larger the circle representing a group, the wider the range it has. For instance, the ester group is large with a range that goes from nearer the horizontal line, a slightly negative or calming quality, to high into the negative range, or much more calming, yin and sedative. Also, the ester group reaches from just over the vertical center to much further right, meaning the range of individual ester molecules are somewhat polar (wet) to a non-polar action (dry), having some water solubility.

The ester circle represents many individual molecules defined as ester compounds by their structure. Each ester compound has a specific electronegativity and polarity that, if charted on the diagram, would appear somewhere in the range represented by the ester circle.

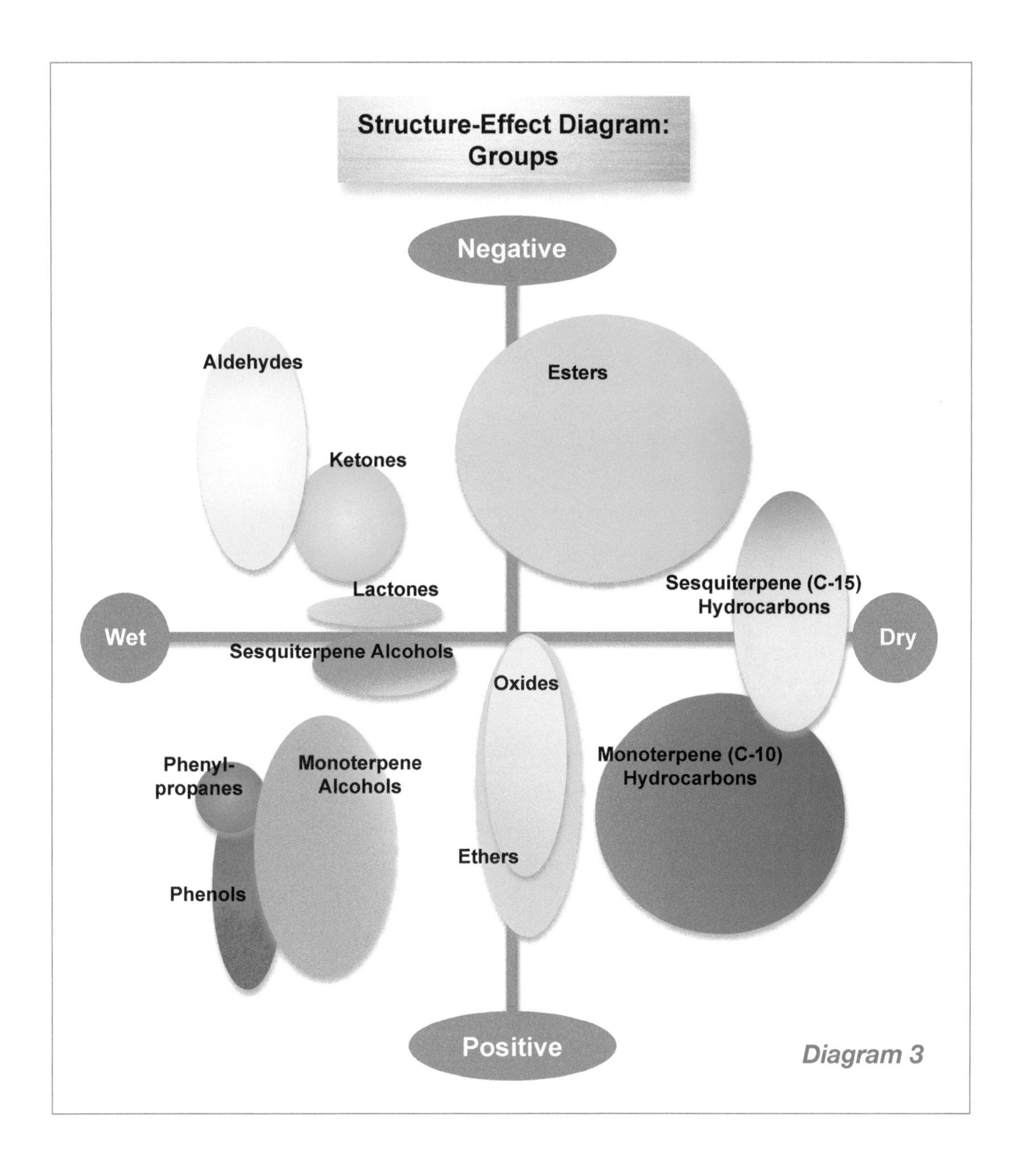
Structure-Effect Diagram:
Groups
Negative
Aldehydes
Esters
Ketones
Lactones
Sesquiterpene (C-15)
Hydrocarbons
Wet
Dry
Sesquiterpene Alcohols
Oxides
Phenyl-
propanes
Monoterpene
Alcohols
Monoterpene (C-10)
Hydrocarbons
Ethers
Phenols
Positive
Diagram 3

Phenols, represented by a red oblong circle shown in the lower left corner, correspond to molecules whose chemical structure defines them as phenols. Phenols have a much smaller range than the ester example. They are positioned lower, or highly electropositive, stimulating and energizing, on the chart and far to the polar left side.

The purpose of this book is to explain how to use the Structure-Effect Diagram to evaluate the therapeutic properties and actions of a family of compounds. This then becomes a guide to understanding the therapeutic properties of the essential oils.

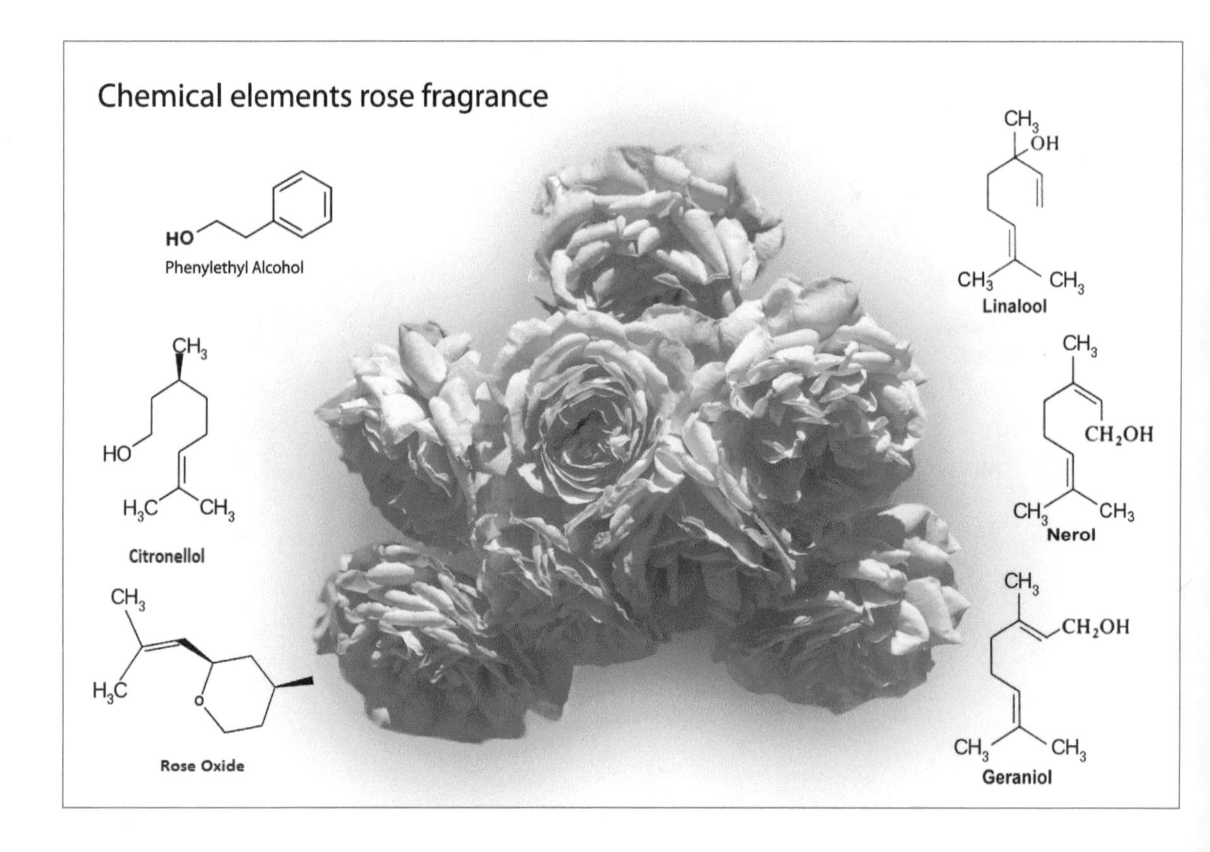

Properties of the Molecular Groups on the Structure-Effect Diagram

Diagram 4 lists some of the therapeutic properties associated with each group.

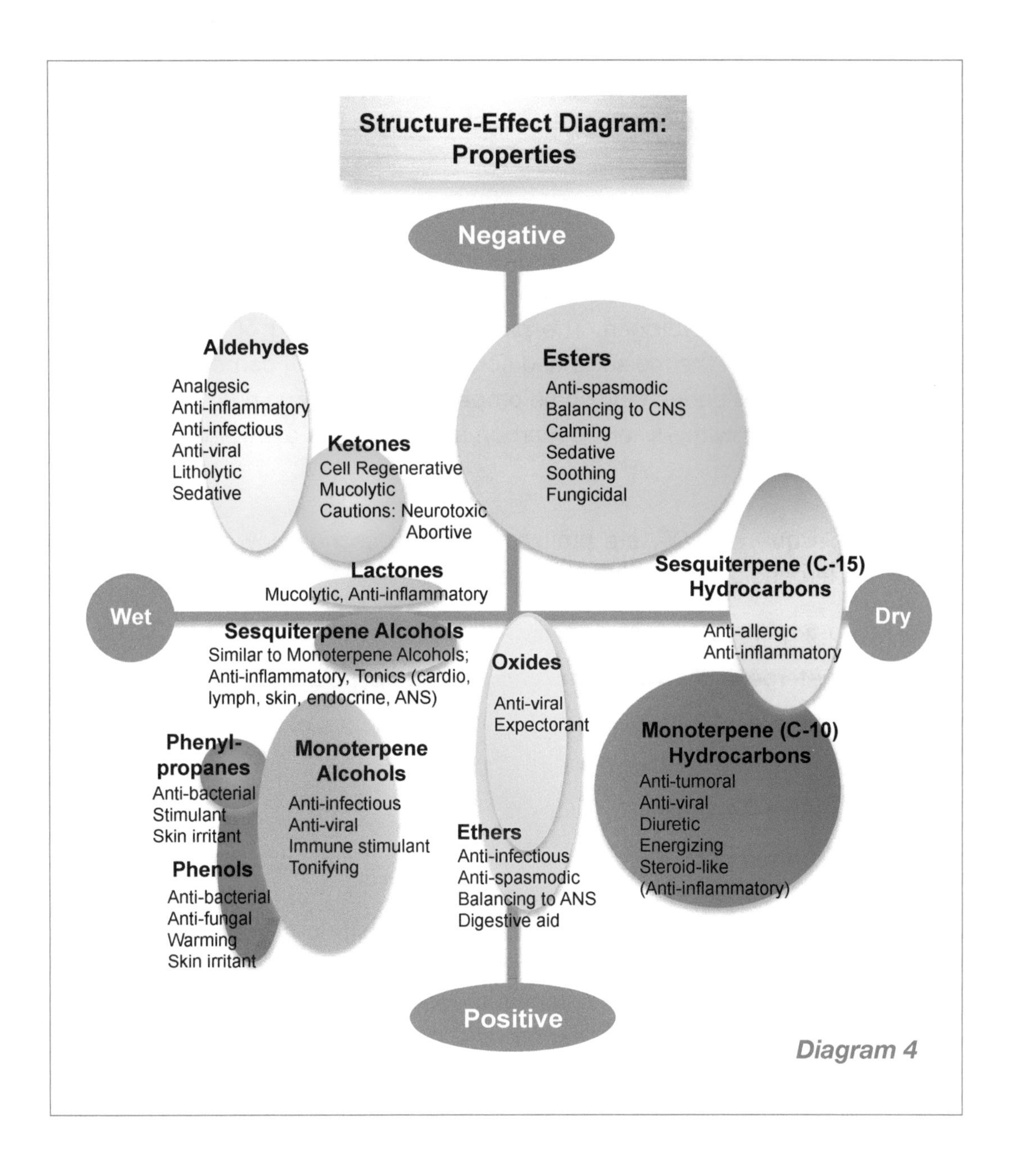

Diagram 4

Defining the Structure-Effect Families

Essential oil compounds are categorized into two main classes: terpenes (terpenoids) and phenyl propane derivatives (phenylpropanoids). These two class structures make up the majority of essential oil compounds. There are a small number of other essential oil compound structures that may not belong to a family included on the diagram. The following paragraphs describe the compound families included on the Structure-Effect Diagram.

Terpenes

Terpene molecules (terpenoids) are the most abundant backbone chemical structure in essential oils. Terpenoids are made up of units composed of 5 carbon atoms and hydrogen. The prefixes in the naming of these groups relate to the size of the molecule and correspond to the 5- carbon unit structure. This helps to understand some properties of the groupings, such as the larger molecules, with 15- or 20- carbon atoms, having a lower evaporation rate, or volatility.

Mono-, *Sesqui-* and *Di-,* are prefixes used to describe how many carbon atoms a terpenoid compound contains:

- *Mono*-terpene is a 10-carbon atom (C10) terpenoid structure.
- *Sesqui*-terpene is a 15-carbon atom (C15) terpenoid structure.
- *Di*-terpene is a 20-carbon atom (C20) terpenoid structure.

Tri- and Tetra-Terpenes

Tri- and tetra-terpenes are too large, at 30 and 40 carbons, to evaporate in the distillation extraction process and, therefore, are not found in essential oils. There is a slight possibility these larger compounds may be found in pressed citrus peel essential oils. These highly antioxidant and anti-inflammatory larger compounds are common in foods, such as beta-carotene (40-carbon atom tetra-terpene) in carrots and oranges. The larger compounds may be found in supercritical carbon dioxide extractions, absolutes and herbal extracts of foods and plants.

The Structure-Effect Compound Groups

The following definitions of the compound groups give a brief explanation of the structure, character and properties of each family. The definitions include a list of essential oils that contain a significant amount of a compound, or compounds, from that group.

Monoterpene (C10) Hydrocarbons

These molecules are the most abundant in essential oils and are composed of a 10-carbon atom chain containing only hydrogen and carbon atoms. Monoterpene hydrocarbons are the most volatile.

Essential Oils: High amounts of monoterpene hydrocarbons are in the needle tree oils (cypress, fir, juniper, pine, spruce), citrus oils (grapefruit, lemon, lime, orange) and in frankincense, greenland moss and opopanax. Compounds from this family are found in almost all essential oils, even if only in low amounts.

Properties: Anti-tumor, antiviral, detoxifying, diuretic, lymphatic support, stimulating, steroid-like (anti-inflammatory), tonic.

Contraindications: Monoterpene hydrocarbons are susceptible to peroxidation, the oxidative degradation of lipids. Oils that contain high amounts of monoterpene hydrocarbons, especially citrus and needle tree oils, may peroxidize easily if stored in light glass, at higher temperatures (above 75° F, 24° C), or stored with lots of head space (space between oil level and bottle neck). Peroxidized essential oils may cause skin irritation.

Sesquiterpene (C15) Hydrocarbons

These molecules are composed of a 15-carbon atom chain and contain only hydrogen and carbon atoms.

Essential Oils: Atlas cedarwood, balsam poplar, black pepper, copaiba, German chamomile, ginger, Helichrysum italicum, katrafay, myrrh, spikenard, vetiver, Virginia cedarwood.

Properties: Anti-inflammatory, anti-allergic, cooling, soothing.

Functional Groups

This next set of compound families are identified by the compound structure (functional groups) attached to a terpene hydrocarbon. The groups are identified and named according to the functional group attached. The functional groups contain an oxygen atom, though sometimes nitrogen or sulfur is attached.

Monoterpene Alcohols

The monoterpene alcohols are defined by an alcohol group (a hydrogen atom bonded to an oxygen atom also called a hydroxy group) attached to a C10 hydrocarbon. These have very broad medicinal properties and are found in higher amounts in many of the herbal essential oils. The monoterpene alcohols are a very beneficial group of compounds, and it's been suggested that, these compounds are the safest of the essential oil molecules, with the exception of menthol.

Essential Oils: Coriander, cilantro, Eucalyptus globulus, Eucalyptus radiata, geranium, lavender, marjoram, Niaouli/MQV, palmarosa, peppermint, rosemary (content dependent upon chemo type), tea tree, and ylang ylang complete.

Properties: Broad medicinal properties, antiseptic, bactericidal, immune modulant and energizing (except linalool, a sedative in lavender, coriander and ylang ylang complete).

Contraindications: Reports show menthol to cause respiratory sensitivity and inflammation in infants, especially when used through nasal passages. Avoid the use of peppermint, a menthol containing oil, on infants and young children under 30 months.

Sesquiterpene Alcohols

These compounds are identified by an alcohol functional group attached to a C15 terpene hydrocarbon.

Essential Oils: Atlas cedarwood, ginger, myrrh, patchouli, balsam poplar, sandalwood (santalum species), vetiver, Virginia cedarwood.

Properties: Anti-inflammatory and antioxidant. Overall the properties are similar to the monoterpene alcohols though this will vary by the individual compound. For example, cedrol in Virginia cedar can regulate autonomic function to reduce "fight or flight" response and (-)alpha-bisabolol is a strong anti-inflammatory found in German chamomile, balsam poplar and Cape chamomile. C15 alcohols are very beneficial, rejuvenating and protective in facial care.

Aldehydes

This functional group is defined by a carbon atom double bonded to an oxygen atom and single bonded to a hydrogen. This, called a carbonyl group, is attached to a terpene compound. When smelling a lemon-like fragrance from an oil, it can be assumed the oil contains aldehydes (usually the lemony compound geranial or citronellal).

Essential Oils: Citronella, cilantro, cumin, Eucalyptus citriodora, geranium, Litsea cubeba, lemongrass, melissa.

Properties: Analgesic, antiseptic, anti-inflammatory, antiviral, hypotensor, sedative.

Contraindications: Can be a skin irritant when used undiluted, mostly related to the lemon-scented aldehydes.

Esters

Esters, the calming and stress-relieving family, are the result of a chemical reaction between an alcohol compound and an acid. For example, linalool, the monoterpene alcohol in lavender, reacts with acetic acid to become the ester linalyl acetate.

> ***Essential Oils:*** Birch, bergamot, Cape chamomile, clary sage, geranium, lavender, mandarin, petitgrain, Roman chamomile, wintergreen.

> ***Properties:*** Anti-inflammatory, anti-spasmodic, anxiolytic (reduce anxiety), balancing to central nervous system, calming, fungicidal.

Ketones

This functional group is defined by an oxygen double bonded to a carbon between two carbons of a terpene structure. Ketones are excellent skin regenerative compounds and are mucolytic. They have the most notorious contraindications and are generally not recommended for casual use in aromatherapy. Helichrysum, Eucalyptus dives and rosemary verbenone type are considered mild ketones, are excellent and highly recommended for skin care, and safe for everyday use.

> ***Essential Oils:*** Atlas cedarwood, Eucalyptus dives, Helichrysum italicum, mugwort, peppermint, rosemary verbenone, sage, spearmint, thuja.

> ***Properties:*** Promotes tissue and cell formation, mucolytic (breaks down hardened mucous).

> ***Contraindications:*** Neuro-toxic, abortive at high dose. Caution using ketone-containing essential oils if prone to seizures; especially high thujone content. Peppermint, with the ketones menthone and pulegone, may be neurotoxic in infant inhalation and is not recommended for children under 30 months, which is also due to its menthol and oxide content.

Sesquiterpene Ketones

This compound group is not represented on the diagram though there are C15 ketones found in some essential oils. Nootkatone, a sesquiterpene ketone, is the compound identified to provide grapefruit its familiar fragrance.

Oxides

When an oxygen atom is integrated into a terpene ring, it's defined as an oxide. The oxide compound is used effectively for respiratory conditions. The oxide 1,8 cineole provides the familiar fragrance and antiviral properties to Eucalyptus globulus and radiata, tea tree and rosemary (cineole type) oil.

Essential Oils: Eucalyptus oils (especially globulus, radiata and smithii), bay laurel, cardamon, tea tree, ravensara, rosemary cineole type, Niaouli/MQV, lavender, peppermint.

Properties: Antiviral, expectorant.

Contraindications: Caution using 1,8 cineole (peppermint, Eucalyptus radiata and Eucalyptus globulus) with infants and young children under 4 years. Simply avoid high dose and direct inhalation.

Phenols

This compound group is defined by an alcohol group bonded to a terpenoid structure with an aromatic ring (a benzene ring). This group is well known for its hot, sensitizing quality and strong antibacterial action.

Essential Oils: Oregano, thyme (thymol type), savory.

Properties: Anti-fungal, anti-parasitic, strong bactericidal, heart tonic, immune stimulant, strengthening, warming.

Contraindications: Skin irritant, possible liver toxin (hepatotoxicity) with prolonged use and high dosage. Thymol is cautioned in conjunction with anti-coagulant drugs as it induces platelet aggregation.

PHENYLPROPANE DERIVATIVES (PHENYLPROPANOIDS)

Phenylpropanoids are the second class of compounds found in essential oils. They are structured within the plant through a different biological pathway than the terpene molecules. There are two phenylpropanoid groups on the Structure-Effect Diagram: ethers and phenylpropanes.

Ether (Mild Phenylpropane Derivatives)

Essential Oils: Anise seed, basil, fennel, nutmeg, tarragon.

Properties: Antibacterial, anti-spasmodic, balances the autonomic nervous system (reduce "fight or flight" response), digestive imbalance (fennel, anise), mental stimulant (basil, nutmeg, tarragon).

Contraindications: Toxic to the nervous system at extreme high dose (primarily nutmeg).

Phenylpropanes (Hot Phenylpropane Derivatives)

Essential Oils: Cinnamon (contains cinnamaldehyde, an aldehyde bonded to a phenylpropanoid compound), clove (with euganol, an alcohol group bonded to a phenylpropanoid).

Properties: Strongly antibacterial and anti-parasitic, similar to phenols; analgesic (clove).

Contraindications: Skin irritant and other similar actions to phenols.

LACTONE AND COUMARIN

These are an interesting family of compounds that are generally found in small amounts in essential oils. A small amount may still have a strong effect in the oils that contain them.

Lactones are produced in the plant through the mevalonate pathway, similar to the terpene compounds. Coumarins are formed as phenylpropane derivatives of cinnamic acid. Furanocoumarins are produced partly through the mevalonate pathway (terpenes) and partly through the phenylpropanoid pathway.

Lactones

Sesquiterpene lactones have diverse healing properties, are anti-inflammatory and have been studied for their ability to stop certain cancers.

Essential Oils: Angelica root, catnip, hibiscus seed, Inula graveolens, jasmine.

Properties: Anti-asthmatic, antibacterial, anti-fungal, anti-inflammatory, anti-tumor, strong mucolytic.

Contraindications: Potential skin sensitivity.

Coumarins

Coumarins are phenylpropane derivatives and have the scent of hay, sweet grass or a freshly mowed lawn. Coumarins contribute to the antispasmodic properties of lavender and khella.

Essential Oils: Khella, lavender, lavender absolute, lavandin absolute, tonka bean absolut.

Properties: Anti-asthmatic and anti-spasmodic (especially noted are khella and lavender), anti-tumor.

Furanocoumarins

Essential Oils: Angelica, bergamot, grapefruit (low percentage in expressed oil), lemon (expressed oil), myrrh.

Properties: Analgesic, antimicrobial, anti-spasmodic, insecticide.

Contraindications: Phototoxic (or photosensitivity) and may cause sunburn and brown spotting—especially the furocoumarin bergapten and psoralen in bergamot, though there is conflicting information as to the level of phototoxicity in lemon, orange and grapefruit. Angelica can be listed as phototoxic, though it, and its compound umbelliferone, is used as sun protection in several over-the-counter sunscreens.

Gas Chromatography/Mass Spectrometry (GC/MS) and a GC Analysis

The Structure-Effect Diagram of an essential oil is based on the oil's molecular content. Gas Chromatography/Mass Spectrometry (GC/MS) is a scientific tool that isolates and identifies the individual chemical compounds of an essential oil. This method uses a scientific devise that creates a graph from the volatile compounds that evaporate from the essential oil being analyzed. The GC displays the individual compounds, shown as peaks and valleys on a graph. From this graph, the testing scientist is able to conclude, using an available data bank, the type and percentage of each compound in the oil. The MS clarifies the structure of each compound.

The benefits of a GC/MS are:

- Identifying the chemical composition of an essential oil
- Identifying the variations in structure of the oils that are distilled from identical botanical plants though from different geographical locations, such as comparing compound structure of a lavender *(Lavandula angustifolia)* from Provence, France and a lavender from Washington, USA
- Identifying contaminants, natural additives or adulterants added to the essential oil following distillation.

A GC/MS is a scientific tool that only a trained specialist, chemist or physicist is able to read. The information contained in a GC analysis is translated into a simpler identification of essential oil composition for the consumer or layperson. This translation is the information available to the consumer that is used to identify the chemical compounds in specific essential oils.

For complete accuracy in breaking down and analyzing an essential oil you own, you would need to have the GC analysis from that specific batch of oil, which can then be translated to an accurate Structure-Effect Diagram. More and more suppliers are offering batch-specific GC analysis. Later in this book we will discuss how to use available GC analysis of essential oils to build Structure-Effect Diagrams.

Marketing Essential Oil Quality and GC/MS

A GC/MS is often promoted by sellers of essential oils as a determination of quality. Though the test, when read by a qualified scientist, may have been used to detect potential adulteration, it does not guarantee or identify the quality or superiority of the essential oil to the consumer.

The quality of an essential oil is not determined by a designated chemical structure shown on a GC. There are other parameters that determine the therapeutic and structure quality of essential oils. It must be understood that quality is subjective. The chemical structure does identify interesting attributes of a tested oil. These attributes may be desired by some and be completely uninteresting to others.

These are some of the qualifiers of a quality essential oil:

- Cultivation (e.g., organic, wild) and geographical location
- Distillation: Determines the final outcome of the essential oil. Distillation is an art that involves timing, pressure, temperature, attention, intuition, and love.
- Authenticity: An authentic essential oil is the true representation of a specific plant identified by the botanical nomenclature.
- Purity: An essential oil(s) that has not had any additives, authentic or natural, following its distillation.
- Complexity: Geographical location, weather conditions, cultivation, distillation, authenticity and purity all come together to determine the complexity of the essential oil.

The words *pure, therapeutic grade* and *high quality* are not standardized terms and are used mainly for marketing purpose as defined by each distributor. These descriptions do not guarantee a quality of the essential oils sold.

To purchase quality essential oil, have an understanding of the art of essential oil distillation. Be aware of compound structure and acquire some ability to read through a GC, which is difficult for most. Acquire the infamous *perfumer's nose* and learn to *feel*—or have a relationship with—the oils. Learn the biological identity and some traits of the plant source. Most important, learn who to trust.

Individual Molecules Listed by Group

A GC/MS analysis is used to identify and list individual constituents contained in the essential oil. To create a Structure-Effect Diagram, compounds are then categorized by the molecular family to which they belong. The following is a list of the more common compounds found in essential oils under the molecular groups to which they belong. For a more detailed list see visit www.jimmharrison.com.

Monoterpene (C10) Hydrocarbons

- d-limonene
- α-pinene/ β -pinene
- p-cymene
- α-ocimene / β-ocimene
- β-myrcene
- α-phellandrene / β-phellandrene
- α-thujene
- sabinene

Sesquiterpene (C15) Hydrocarbons

- caryophyllene
- chamazulene
- farnesene
- elemene
- bisabolene
- humulene
- cadinene
- bergamotene

Monoterpene Alcohols

- borneol
- geraniol
- linalool
- menthol
- nerol
- terpineol
- terpinen-4-ol
- thujanol

Sesquiterpene Alcohols

- (-)α-bisabolol
- cedrol
- santalol
- nerolidol
- farnesol
- viridiflorol

Aldehydes

- citral
- citronellal
- cuminal
- decanal
- geranial (The E-isomer of citral)
- neral (The Z-isomer of citral)
- octanal

Esters

- benzyl acetate
- citronellyl formate
- geranyl acetate
- geranyl tigate
- isobutyl angelate
- linalyl acetate
- N-methyl anthranilate
- neryl acetate

Monoterpene Ketones

- camphor
- cedrone
- menthone
- piperitone
- thujone
- verbenone
- r-carvone
- s-carvone

Phenols

- carvacrol
- thymol

Oxides

- 1,8 cineole
- caryophyllene oxide
- rose oxide

Ethers

- anethole
- methyl chavicol (estragole)
- methyl euganol
- myristicin

Hot Phenylpropanoids

- Cinnamaldehyde
- Euganol

ⓘ FOR YOUR INFORMATION

Cinnamaldehyde is Not a Terpene Aldehyde

Euganol does not belong to the phenol group of the structure diagram and cinnamaldehyde is not in the aldehyde group. These two compounds, and similar compounds, are often improperly categorized in these groups by online sellers, essential oil brands and even by respected scientists. Though cinnamaldehyde has an aldehyde functional group, making it correctly an aldehyde type, it is not a terpene compound.

The structure of both of these compounds begins with a phenylpropane backbone structure. The aldehyde and the phenol groups on the diagram are based on terpene backbone structures. For euganol the improper category phenol is not a big issue, since phenols have similar properties to the hot phenylpropanoids. Categorizing cinnamaldehyde as an aldehyde could lead to serious misuse if matching the use of cinnamon to the properties of the aldehydes. For intelligent and safe classification, know that cinnamaldehyde and euganol are hot phenylpropanoids.

Fragonard
PARFUMEUR
Concerto

PART 2

How to Create a Structure-Effect Diagram

Essential oils have properties that are known:

- From their history of use and popular knowledge
- From the repeated successful treatment of identified conditions
- Through scientific evaluation and research

When you don't know what the properties are of an essential oil, the most common way to find that information is to look it up in trusted resources.

The Structure-Effect Diagram provides an alternate guide to the potential properties of an essential oil and can be used to select oils for specific conditions. The oil's potential properties become apparent with a visual structure-effect representation. This is especially beneficial if that oil is unfamiliar.

There are several possible ways to construct a visual representation of an essential oil according to its molecular structure. In the following diagram for Orange *(Citrus sinensis)* you can see which groups are found in this oil. They are represented by a percentage range contained for each group. If important to the understanding of the therapeutic action of the oil, a specific compound from the group is listed, particularly if the single compound is contained at a significant amount or if its action has a marked value to the overall action of the oil.

Sweet Orange Profile

In Diagram 5, the compound limonene is listed as a monoterpene hydrocarbon contained in orange essential oil. Limonene is a main component in many essential oils, especially citrus oils. The monoterpenes are diagrammed here to be up to 96%. This does not mean limonene alone is up to 96%. All listed percentages are an estimated evaluation of the known contents of orange essential oils. From this diagram and the high amount of monoterpene hydrocarbons, especially limonene, it can be determined that orange essential oil will have detoxifying and drying properties. The diagram also shows, from the lesser amounts of the other compound families, how orange gets its distinctive properties, qualities, fragrance and character.

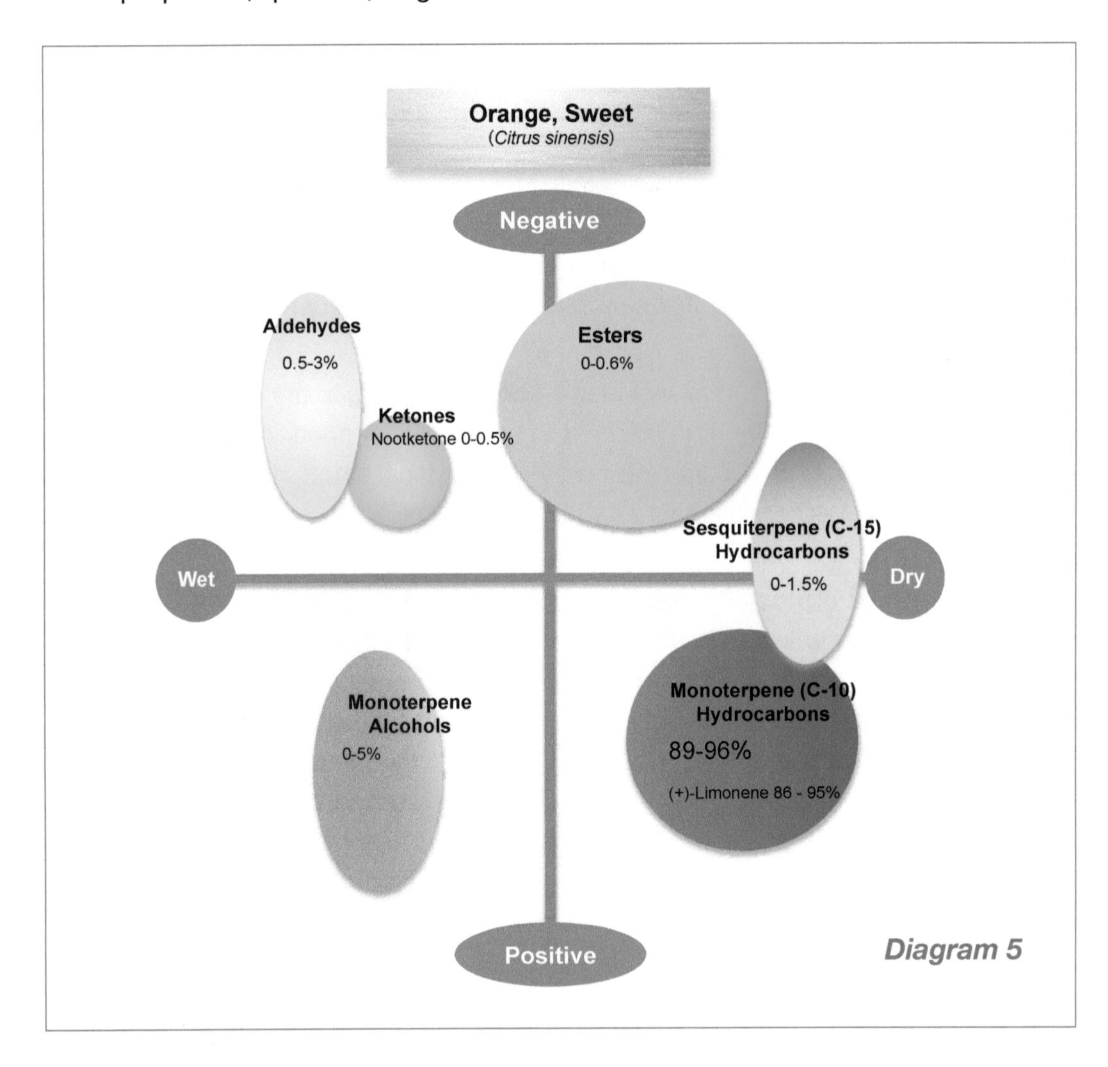

Diagram 5

Cinnamon Bark Profile

Cinnamon bark essential oil contains (E)-cinnamaldehyde, a phenylpropanoid, in the highest percentage (63-75% range). In Diagram 6 it can be seen there is another notable phenylpropanoid compound listed, euganol. It is understood that most of the therapeutic action is coming from this hot and highly antimicrobial family. Note the interesting natural balance to its structure by the content of cooling, anti-inflammatory sesquiterepene hydrocarbons that temper the hot phenylpropanoids. Cinnamon may at rare times have a high cinnamyl acetate ester content, up to 13% as shown in the diagram's ester range. This quality results in a much "softer" cinnamon essential oil with the hot phenylpropanoids at a lower percentage.

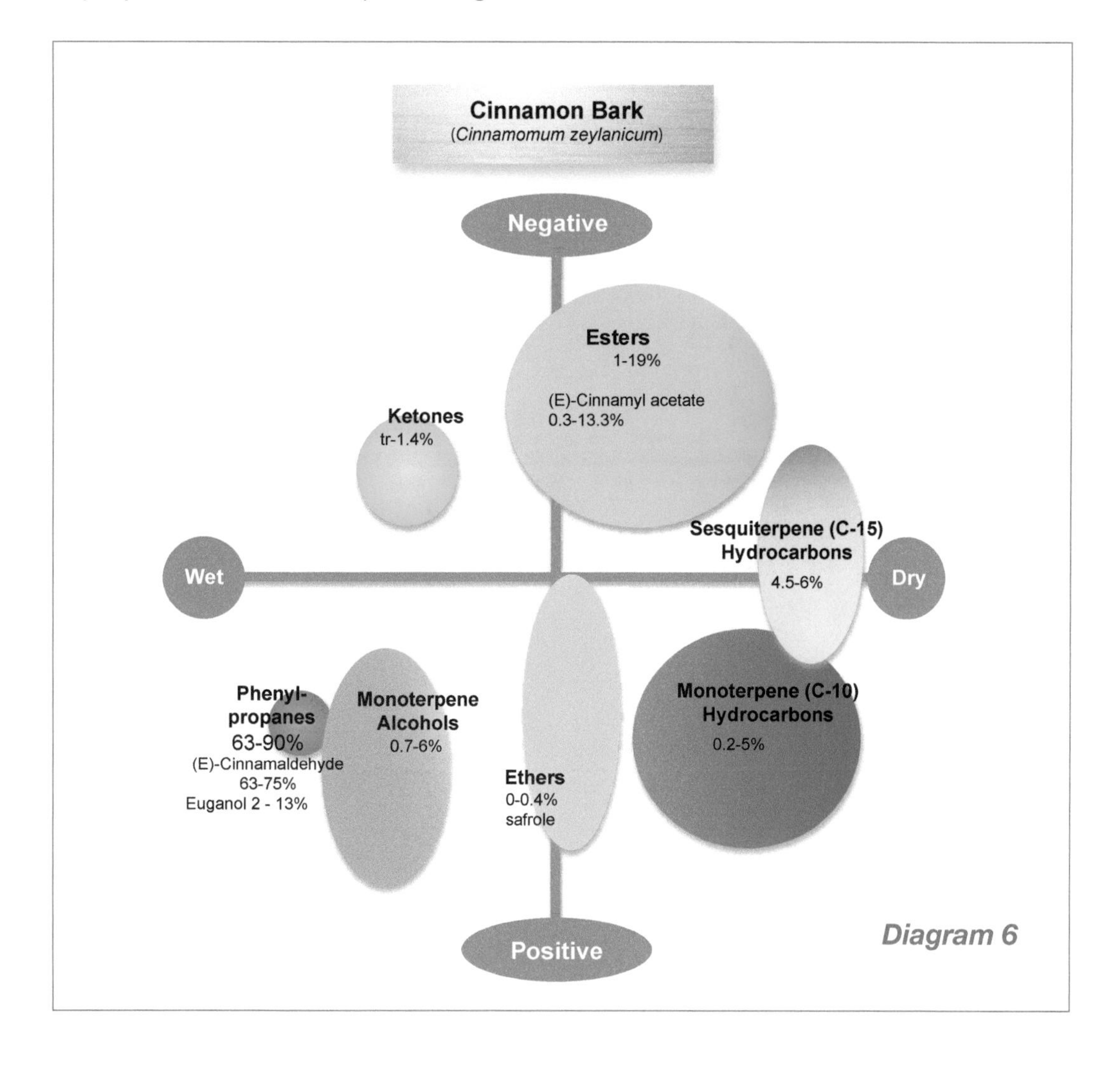

Diagram 6

Lavender Profile

Diagram 7 represents a common composition of lavender *(lavandula angustifolia).* In lavender essential oil the linalyl acetate content is between 33–40% with a linalool content within the same range. The linalyl acetate and linalool content are together as much as 75–80% of the lavender oil. There could be well over 100 compounds in a skillfully distilled lavender essential oil. Lavender has a diverse complex synergy of the compound families which is why it is such a versatile essential oil.

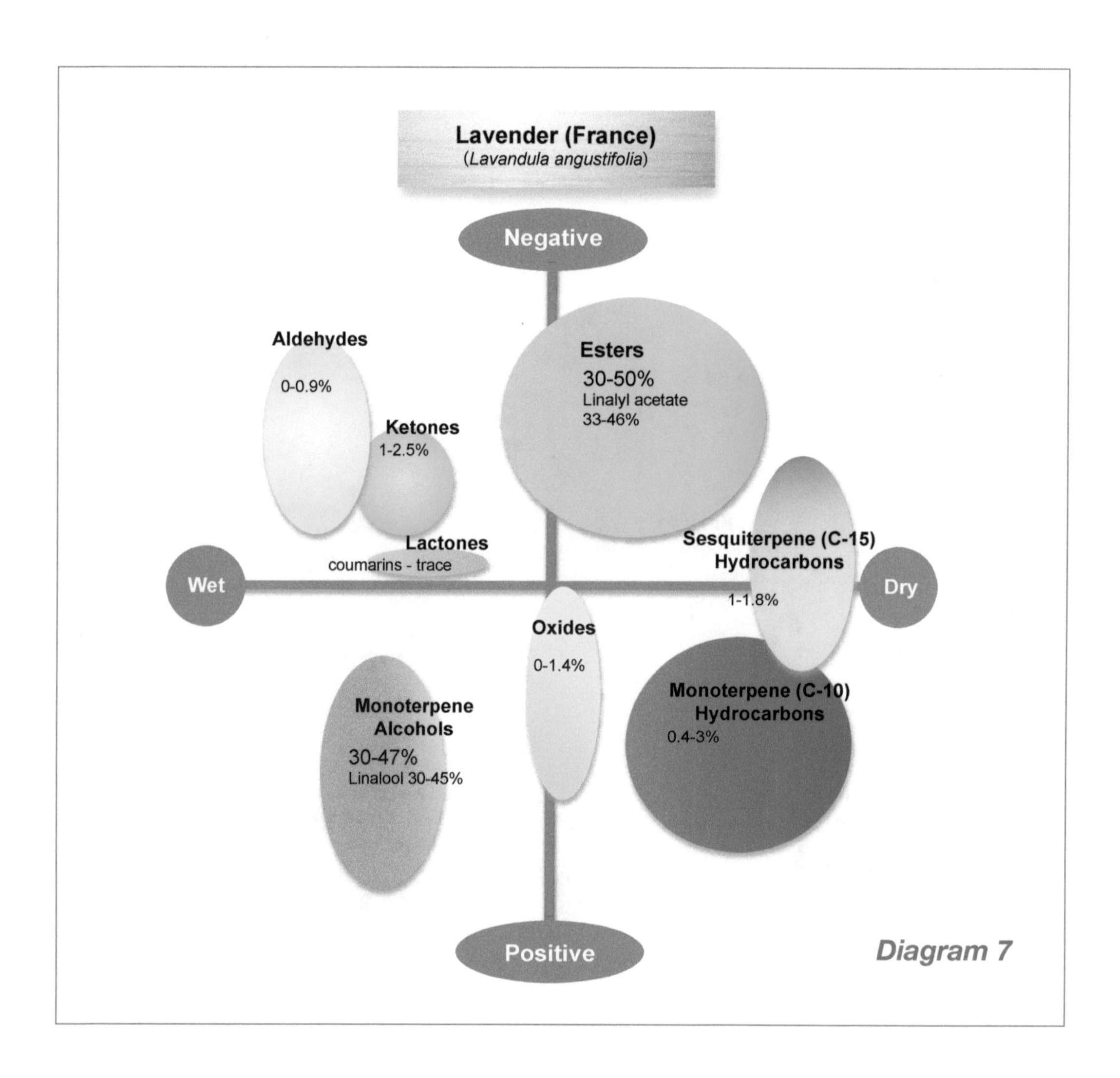

Diagram 7

Creating an Essential Oil Structure-Effect Diagram: Peppermint *(Mentha piperita)*

Following are the steps used to create a Structure-Effect Diagram for an essential oil. The oil used in this example is peppermint. It is important to know specifically which plant—its botanical name rather than the common name—the essential oil is distilled or pressed from. This is called specified botanical origin. Plants from different species have their own unique structure and range of compounds. For more accuracy it's also helpful to know the geographical location the plant is grown. The botanical origin of peppermint in this example is *Mentha piperita* from France.

To create a Structure-Effect Diagram:

1. Define the botanical origin and geographical location of the oil.
2. List the compounds of the essential oil and at what percentage they occur.
3. Organize the compounds by group.
4. Provide a total percentage or range for each group.
5. Fill in the Structure-Effect Diagram with the total percentage, or range, of each group contained in the essential oil.

STEP ONE

Identify and List the Compound Groups and Individual Compounds Contained in the Oil

When the exact, batch-specific, GC/MS or compound analysis of an essential oil is not available, an analysis of a comparable essential oil (identical botanical origin and location) can be used to create a Structure-Effect Diagram. The analysis used will have variations on how the compound content is listed, depending on your source. The analysis may be laid out with each compound listed with an exact percentage amount. Some lists found will be compiled from several analysis of essential oils distilled from similar botanical origin and location, giving a percentage or percentage range based on the varied results. The resulting lists can be written as < or > amounts, such as <10%, or in a percentage range, like the following peppermint example, using ranges such as 15–27%. From these lists, a best guess can be made as to what range to chart for the Structure-Effect Diagram.

A complete and accurate GC analysis from a specified oil will list the compounds in the order they appear. Those that easily evaporate will appear first. The following analysis for peppermint will order the compounds from the most to the least amount in the oil.

A GC analysis may show hundreds of compounds. Most compound lists made available to the consumer will already be broken down to the compounds contained at significant amounts or edited for simplification. A range is used when many samples are compared. Though the range may be wide, this still gives a good idea of what an oil's composition could be. The only way for complete accuracy is to have the exact GC analysis of the essential oil, which isn't necessary for the therapeutic purposes proposed in this text.

The total percentage of each individual compound will be listed followed by a total percentage range. Some compounds or groups will have an amount or range starting with"tr," referring to the compound or group having trace amounts. Trace amounts do contribute to the overall character of the essential oil.

Following is the list of compounds commonly found in French peppermint *(Mentha piperita)** essential oil.

- Menthol 35–46%
- Menthone 15–27%
- Isomenthone 3–8%
- Menthyl acetate 2–10%
- Neomenthol 2.5–6%
- 1,8-Cineole 2–6%
- Limonene 1–3%
- Menthofuran 1–6%
- Pulegone 0.5–3%
- Germacrene D tr–4.4%
- beta-Pinene 0.6–2%
- trans-Sabinene hydrate (also trans-4-thujanol) 0.2–2.4%
- Terpinen-4-ol 0.10–5%

- beta-Ylangene 0–0.6%
- beta-Caryophyllene 0.1–2.8%
- Sabinene 0–0.4%
- alpha-Thujene 0–0.35%
- beta-Pinene 0.35–2%
- Piperitone 0–1.3%
- gamma-Terpinene 0–0.25%
- (E)-beta-Ocimene 0–0.25%
- beta-Bourbonene 0–0.2%

*More compounds are found in peppermint oil. This list is edited for simplicity.

When working with a single compound range rather than having a total group range, as in this example, it will be necessary to come up with a reasonable percentage range that represents the total for each group. Some compounds may be 35–46%, another may be 2.5–6% and still another compound 0.1–5%, all within the same group (C10 alcohol). The range for the group would not start from the lowest number to the highest, as in 0.1–46%. The total is a possible combination of all, and when using a range, it is unknown the exact totals of each compound and how they will combine in the essential oil.

Since the compound contained at the highest amount starts at 35%, we know that this group will have at least 35%. Looking at the low percentages, we will assume that at least 3% can be added. A best guess for this example group would be in a range of 38–49%. Not at all exact or scientific, but does provide a workable range for the oil. Do not expect totals to add up to 100%. There is a fair amount of unknown and unlisted compounds that make up the remainder. If a total seems to go beyond 100% it is due to the inaccuracies of using a range.

By identifying and categorizing each compound, by either known identification or research, a list by group can be made as shown:

Compound Family	Individual Compounds	Percentage Range
Monoterpene Alcohol	Menthol 35–46%, Neomenthol 2.5–6%, trans-4-thujanol 0.2–2.4%, Terpinen-4-ol 0.1–5%	38–49%
Ketones	Menthone 15–27%, Pulegone 0.5–3% Isomenthone 3–8%, Piperitone 0–1.3%	25–34%
Esters	Menthyl acetate 2–10%	2–10%
Oxide	1,8-Cineole 2–6%	2–6%
Monoterpene Hydrocarbons	Limonene 1–3%, beta-Pinene 0.6–2%, gamma-Terpinene 0–0.25%, Sabinene 0–0.4%, alpha-Thujene 0–0.35%, (E)-beta-Ocimene 0–0.25%	3–5%
Sesquiterpene Hydrocarbons	Germacrene D tr–4.4%, beta-Ylangene 0–0.6%, beta-Caryophyllene 0.1–2.8%, beta-Bourbonene 0–0.2%	1–6%
C10 Furans (Though not on structure diagram - List under lactones)	Menthofuran 1–6%	1–6%

How to Identify What Family a Compound Belongs To

For many available GC/MS evaluations, research may be necessary to find the family to which the compound belongs. This information is available from an online search or other resource that offers chemical compound evaluation with group identification. A list of compounds organized by family group can be found at **www.jimmharrison.com.**

A starting guide to identifying the chemical compound is to look at the structure or "formula" of the chemical compound. Your online search for this information may include chemical resource websites. If the compound formula contains C10, it is a monoterpene, a C15 is a sesquiterpene, and C20 is a diterpene. If there is an O (oxygen) in the formula, it belongs to a functional group—which has to be further identified. If there is not an O in the formula it is either a monoterpene (C10) hydrocarbon or a sesquiterpene (C15) hydrocarbon.

Using an online resource, you may find the formula of a chemical compound, such as the formula for linalool, $C_{10}H_{18}O$. This shows that it is a monoterpene compound (C10) and that it has oxygen (O), but does not completely identify by name the family to which the compound belongs.

As a general guide, the suffix of the compound name can help identify its family group. For example, if the compound has the suffix -ol, it belongs to one of the terpene alcohol groups (with the exception of phenols, thymol and carvacrol; and euganol, a phenylpropanoid). The suffix -ol refers to the functional group alcohol. The amount of carbons in the formula will tell you whether it's a mono-, sesqui- or di- terpene alcohol. A compound with 15 carbon atoms, C15, and an -ol suffix is a sesquiterpene alcohol, such as bisabolol with a formula of $C_{15}H_{26}O$.

The common suffix for essential oil compounds are:

-ene refers to terpene hydrocarbons
-ol refers to alcohols
-al refers to aldehydes
-one refers to ketones
-ate refers to esters

If the alcohol functional group, -ol, bonds to a benzene ring or a phenylpropanoid it is either categorized in the phenol group (thymol and carvacrol) or phenylpropanoid group (euganol). The aldehyde group, -al, with a bond to a phenylpropanoid is categorized in the phenylpropanoid group (cinnamaldehyde).

STEP TWO

Place the Percentage on a Structure-Effect Diagram

The next step is to place the percentage range for each group onto a Structure-Effect Diagram. In Diagram 8, the compound groups that are not represented in the essential oil have been removed to give a better visual representation of the overall structure.

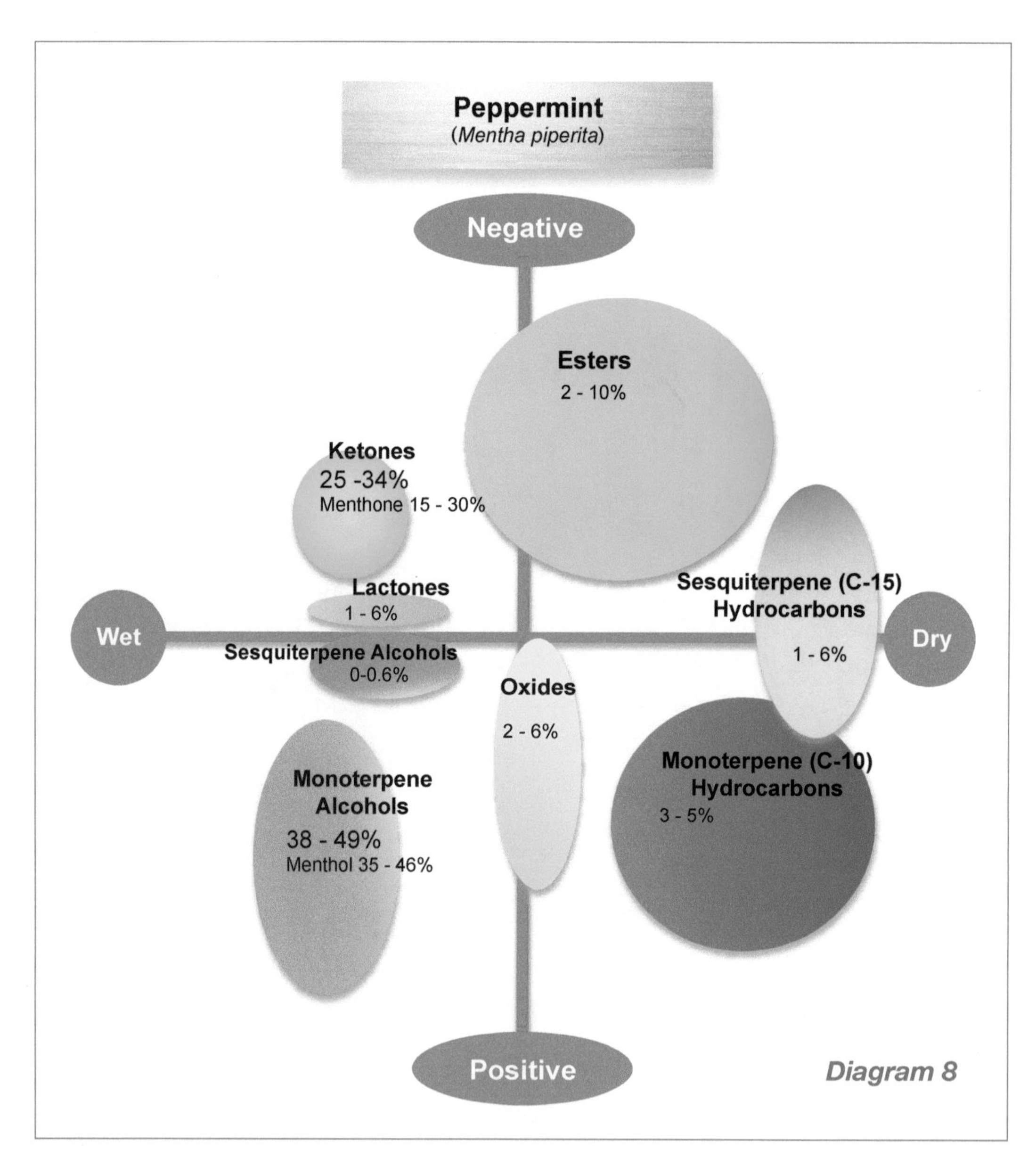

Diagram 8

STEP THREE

Evaluate Potential Therapeutic Action of the Oil

The completed Structure-Effect Diagram for peppermint essential oil indicates potential therapeutic uses:

- Menthol and the other monoterpene alcohols are predicted to provide energizing properties and a synergy for overall wellness of the body.
- Menthol is a compound that has analgesic properties and gives that icy-cool feel when used for headaches.
- Use of peppermint essential oil for respiratory issues is reasonable due to the expectorant action of the oxide and the high ketone content, a compound group with mucolytic properties (breaks down hardened mucous).
- The synergy of esters, sesquiterpene hydrocarbons, and monoterpene hydrocarbons, along with furans (categorized within the lactone group on the chart), show peppermint to have potential diverse action on the nervous system, are anti-inflammatory, anti-parasitic, and promote overall health.
- It is the synergy of therapeutic compounds that provide the well-known gastrointestinal properties of peppermint.
- Contraindications and cautions (not recommended for children under 30 months old) are made evident by the content of menthol, the oxide 1,8-cineole and the ketones.

Creating a Structure-Effect Diagram from a GC/MS for a Batch-Specific Essential Oil

It's possible to create a diagram with a more exact percentage for the groups and the individual compounds. This is only accurate if you have the exact GC analysis for the oil being identified on the structure diagram. A GC for a specific oil will be identified with a batch number and supplied by either the producer, the distributer of the oil or by having the oil sent out for testing. If the analysis from the identified oil is not available, it's best to work from a range, as shown in the peppermint example.

The following table is a GC analysis provided by a Canadian distiller for a batch-specific balsam poplar *(Populus balsamifera)*. An exact Structure-Effect Diagram can be created for this oil from the analysis and compounds listed. The properties can then be determined with an evaluation of the Structure-Effect Diagram of this essential oil.

GC ANALYSIS OF CANADIAN BALSAM POPLAR (FROM THE DISTILLER)

Compound	Percentage	Compound Group
α–Pinene	0.10	Monoterpene Hydrocarbon
Myrcene	0.06	Monoterpene Hydrocarbon
para-Cymene	0.07	Monoterpene Aromatic
Limonene	1.07	Monoterpene Hydrocarbon
γ-Terpinene	0.12	Monoterpene Hydrocarbon
Terpinen-4-ol	0.18	Monoterpene Alcohol
α-Ylangene	0.34	Sesquiterpene Hydrocarbon
α-Copaene	0.47	Sesquiterpene Hydrocarbon
2-epi-α-Cedrene	0.19	Sesquiterpene Hydrocarbon
α-Cedrene	0.51	Sesquiterpene Hydrocarbon

Compound	Percentage	Compound Group
β-Caryophyllene	0.60	Sesquiterpene Hydrocarbon
trans-α-Bergamotene	2.28	Sesquiterpene Hydrocarbon
6,9-Guaiadiene	0.49	Sesquiterpene Hydrocarbon
α-Humulene	0.46	Sesquiterpene Hydrocarbon
Alloaromadendrene	0.51	Sesquiterpene Hydrocarbon
trans-β-Farnesene	2.22	Sesquiterpene Hydrocarbon
γ-Gurjunene	0.31	Sesquiterpene Hydrocarbon
trans-Cadina-1(6), 4-diene	0.50	Sesquiterpene Hydrocarbon
γ-Muurolene	0.91	Sesquiterpene Hydrocarbon
α-Amorphene	4.61	Sesquiterpene Hydrocarbon
β-Selinene	1.67	Sesquiterpene Hydrocarbon
δ-Selinene	0.63	Sesquiterpene Hydrocarbon

Compound	Percentage	Compound Group
α-Selinene	1.50	Sesquiterpene Hydrocarbon
Epizonarene	1.43	Sesquiterpene Hydrocarbon
α-Muurolene	1.67	Sesquiterpene Hydrocarbon
δ-Amorphene	1.63	Sesquiterpene Hydrocarbon
β-Bisabolene	0.68	Sesquiterpene Hydrocarbon
β-Curcumene	4.45	Sesquiterpene Hydrocarbon
7-epi-α-Selinene	0.52	Sesquiterpene Hydrocarbon
δ-Cadinene	4.86	Sesquiterpene Hydrocarbon
trans-Cadina-1,4-diene	1.12	Sesquiterpene Hydrocarbon
trans-γ-Bisabolene	1.40	Sesquiterpene Hydrocarbon
trans-α-Bisabolene	1.22	Sesquiterpene Hydrocarbon
Selina-3,7(11)-diene	0.60	Sesquiterpene Hydrocarbon
trans-Nerolidol	2.87	Linear Sesquiterpene Alcohol

Compound	Percentage	Compound Group
Guaiol	0.19	Sesquiterpene Alcohol
Fokienol	0.22	Linear Sesquiterpene Alcohol
1,10-di-epi-Cubenol	1.05	Sesquiterpene Alcohol
γ-Eudesmol	4.03	Sesquiterpene Alcohol
τ-Cadibol	1.48	Sesquiterpene Alcohol
β-Eudesmol	5.78	Sesquiterpene Alcohol
α-Eudesmol	4.85	Sesquiterpene Alcohol
α-Cadinol	1.53	Sesquiterpene Alcohol
epi-β-Bisabolol	0.73	Sesquiterpene Alcohol
β-Bisabolol	1.52	Sesquiterpene Alcohol
α-Bisabolol	31.91	Sesquiterpene Alcohol
Tricosane	0.17	Linear Alkane

Using the percentage of each compound, a total can be created for each family group. The following table is a total percentages for each family group of the Balsam Poplar. These exact totals can then be diagrammed on a Structure-Effect Diagram, as shown in Diagram 9.

Compound	Percentage Total
Monoterpene Hydrocarbon	1.42
Monoterpene Alcohol	0.18
Sesquiterpene Hydrocarbon	37.78
Sesquiterpene Alcohol	56.16
Linear Alkane (Alkanes are not grouped on diagram)	0.17

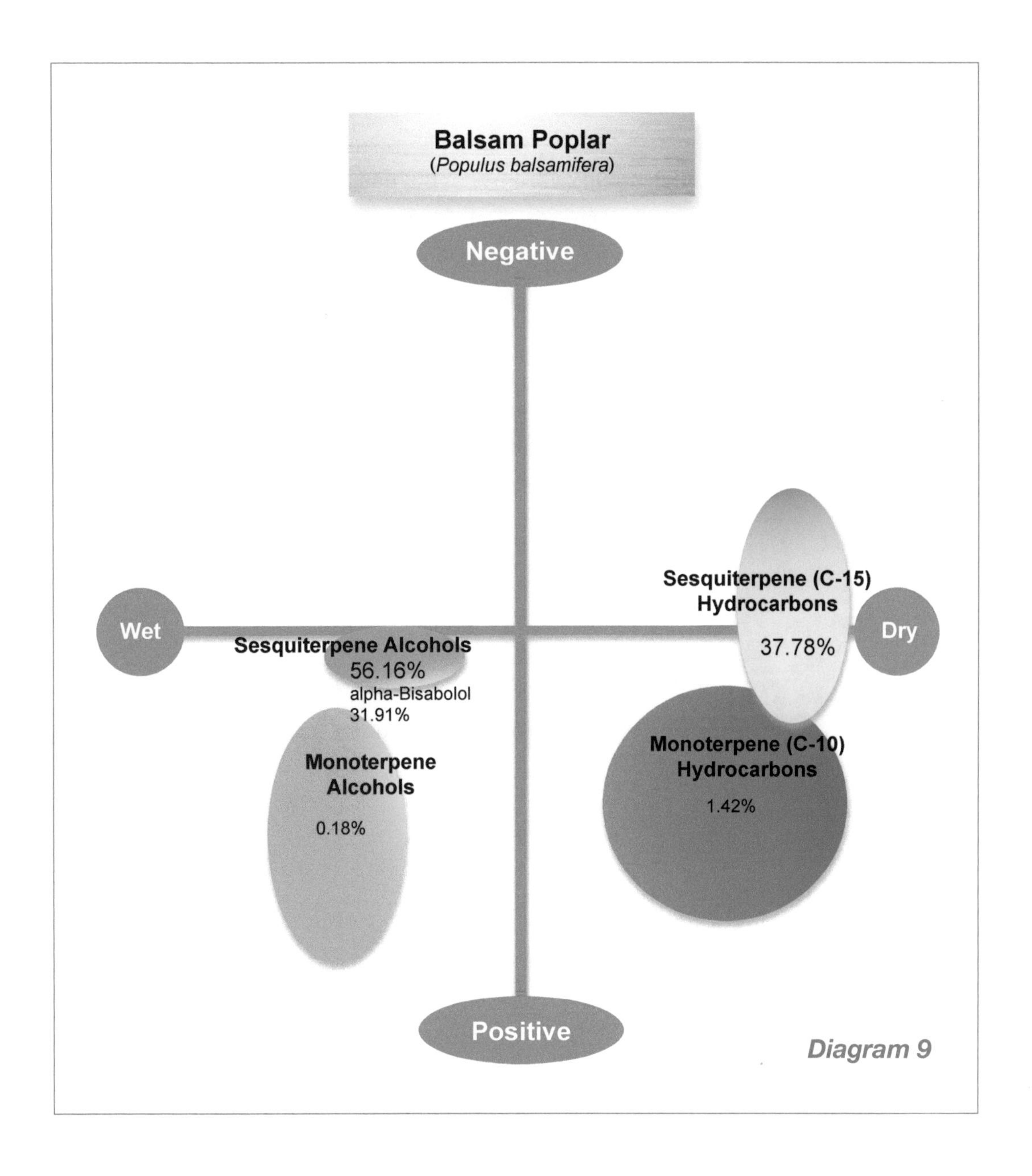
Balsam Poplar
(Populus balsamifera)
Negative
Wet
Dry
Positive
Sesquiterpene Alcohols
56.16%
alpha-Bisabolol
31.91%
Monoterpene
Alcohols
0.18%
Sesquiterpene (C-15)
Hydrocarbons
37.78%
Monoterpene (C-10)
Hydrocarbons
1.42%
Diagram 9

Structure-Effect Diagram for Balsam Poplar, Canada

(Populus balsamifera)

Using the Chemical Structure to Anticipate Potential Actions of Balsam Poplar *(Populus balsamifera)*

- Balsam poplar is composed mainly of sesquiterpene compounds (alcohols and hydrocarbons), making up over 90% of the essential oil.
- The synergistic mix of sesquiterpene hydrocarbons and sesquiterpene alcohols would suggest potent anti-inflammatory action for this oil.
- The sesquiterpene alcohol, α-Bisabolol, at 31.91%, is a well-documented anti-inflammatory compound.
- Overall, from this structure diagram, it can be anticipated that this essential oil would be highly effective for use in any inflammatory condition and likely a very good skin conditioning and regenerative essential oil.

Further Adventures in Creating a Structure-Effect Diagram

When researching an essential oil's structure from available information, there may be inconsistencies, a limited amount of information, or overwhelming scientific data. The more that is learned about the chemistry of essential oils, the easier it will be to understand available information.

Refer to many sources to get the most accurate compound percentage for the essential oil being researched. Geographical location, the vintage year of distillation and the specific distiller determine the very individual details of an essential oil's composition.

Not all lavenders are the same; not all peppermint, or bay laurel or any other oil will have the exact same structure unless they are from the same distillation. This sets up another reminder that the chemistry and Structure-Effect Diagram are only guides to the use of essential oils.

Looking through available GC analysis of essential oils will demonstrate the extreme diversity in essential oil structure. For example, in a study titled *Volatile Composition and Antimicrobial Activity of Twenty Commercial Frankincense Essential Oil Samples* there are twenty frankincense essential oil samples analyzed for their composition. The content of the nine *Boswellia carteri* samples had a range for α-pinene from 4.8 to 40.4%. Other compounds such as limonene and α-thujene had similar wide ranges. These are realistic differences possible in essential oils from varied distillations and geographical locations.

Further review brings up many questions that are important to be aware of when evaluating studies or other informational sources. In this frankincense paper it is stated that the samples tested were "commercial" oils "purchased at various herbal shops or pharmacies." The source of the samples is questionable in regard to potential adulteration, guarantee of plant authenticity, and quality and method of the distillation. These are all important facts when accuracy of source and quality is desired. The research journey may reveal interesting discoveries regarding the essential oils of interest. The more you review, the more you will understand and see how the essential oil adventure is not linear and sometimes asks more questions than are answered.

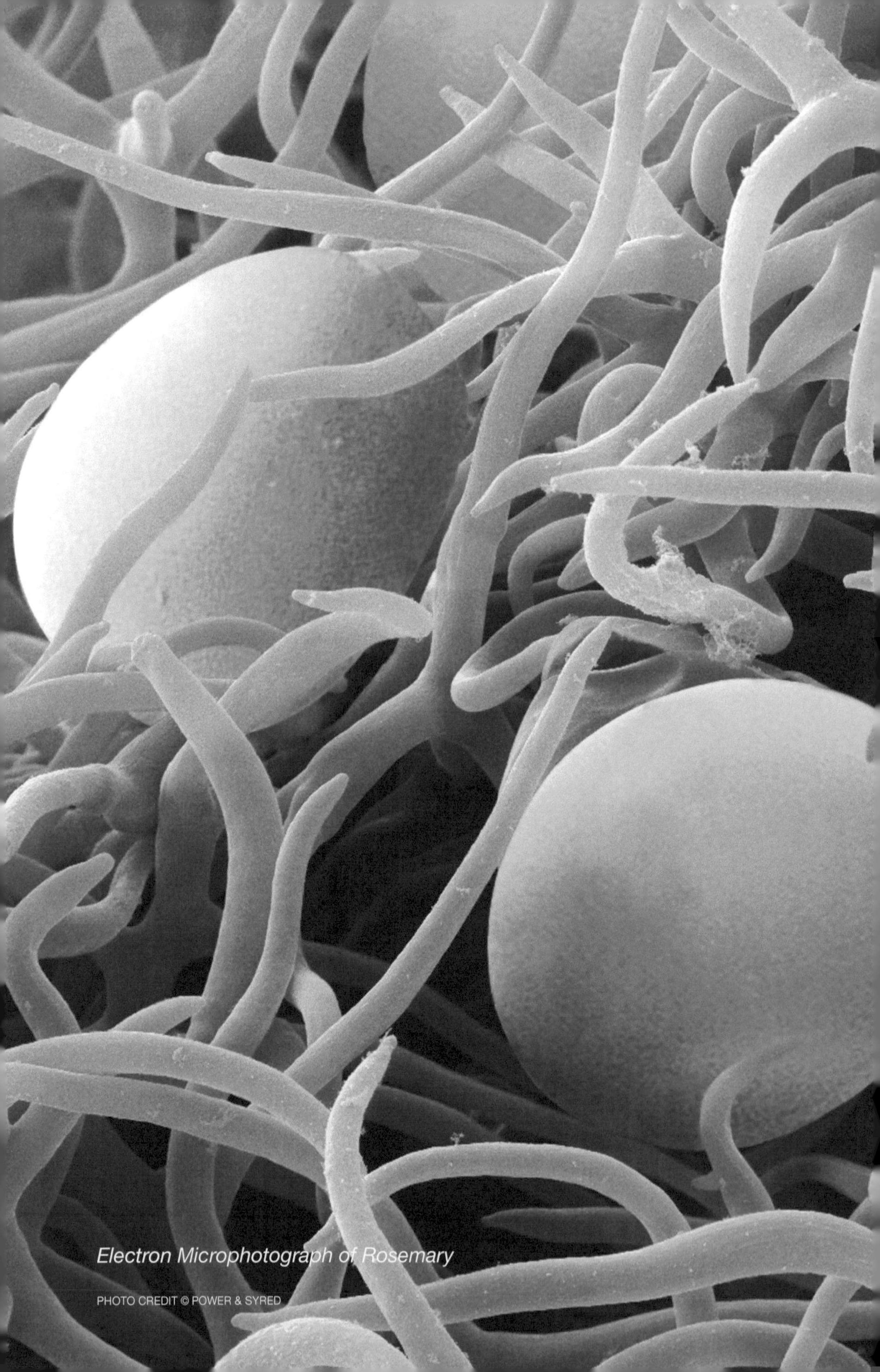

Electron Microphotograph of Rosemary

PART 3

Using the Structure-Effect Diagram for Therapeutic Focus

Creating a Structure-Effect Diagram will:

- Provide a better understanding of the properties of each essential oil
- Be a guide to formulating more effective and specific therapeutic essential oil blends
- Provide direction to alternate essential oil choices having similar therapeutic properties to an oil(s) that may not be available

The Structure-Effect Diagram is primarily used to make essential oil choices for a therapeutic outcome. For instance, if the desired outcome is to relieve an itchy, inflamed skin condition, the diagram can be used to select essential oils that contain the compounds with properties that would best address this condition. Following are instructions on how to select essential oils for therapeutic action and results using the Structure-Effect Diagram.

STEP ONE

Decide your therapeutic focus

Just like programing your GPS, a destination or desired outcome needs to be defined. Desired outcome could be to relieve digestive issues, calm a busy mind, promote restful sleep, heal a wound, prevent a flu, or reverse the skin's sun damage. This is referred to as the therapeutic focus or goal, and the purpose of essential oil selection(s). The following example will use relieving a headache as the therapeutic goal.

In any condition or symptom, such as a headache, there are other health factors or causes that need to be uncovered. Evaluation, through a structured or informal consultation, will assist in discovering potential causes of the headache. *Please be aware of any state and federal regulations regarding diagnosing illness and the use of essential oils.*

A holistic process is described in Jimm Harrison's *Aromatherapy* (Chapters 1 and 13).

Headaches can be caused by many different factors including diet, environment, allergies, hormone imbalance and stress. For this example, stress and nervous system related conditions will be identified as the potential cause(s) of the headache. Related symptoms and causes are included with the therapeutic focus as listed here:

- headache
- stress
- anxiety
- inflammation
- heart palpitations
- insomnia
- muscle tension (neck and shoulders)

STEP TWO

List the properties that best address the therapeutic focus, the symptoms and the possible related causes

It's easy to list properties for some of the selected conditions simply by putting an anti- in front of the condition, such as anti-stress, anti-anxiety and anti-inflammatory. Insomnia would call for sedating properties as well as properties that raise parasympathetic nervous system function—rest and digest mode (see FYI: Fight-or-Flight Response, page 50). Analgesic properties would be included for the headache pain.

The properties for this example are:

- anti-anxiety
- anti-inflammatory
- anti-stress
- analgesic (pain relief)
- sedative
- stimulate parasympathetic function (reduce fight or flight response)

STEP THREE

Select the molecular compound group(s) that provides the properties needed

The properties of the structure groups become more familiar with repeated use of Structure-Effect Diagrams. Until then keep resources accessible when choosing the compound groups for the therapeutic goal. Refer to The Structure-Effect Diagram: Properties on page 11 and the description of the compound families on pages 12-19.

The chemical groups with the desired properties for this example (headache) are:

Anti-inflammatory

- sesquiterpene hydrocarbons
- sesquiterpene alcohols
- aldehydes
- oxides

Anti-stress

- esters
- aldehydes
- ethers

Analgesic

- aldehydes
- sesquiterpene hydrocarbons
- ethers
- esters
- euganol (a phenylpropanoid)
- oxides
- menthol (a monoterpene alcohol)

Stimulate parasympathetic function (reduce fight-or-flight response)

- ethers
- esters
- cedrol (a sesquiterpene alcohol found in Virginia cedarwood and in some cypress oils)

ⓘ FOR YOUR INFORMATION

The Fight-or-Flight Response

Nervous tension and anxiety may raise sympathetic nervous function's fight-or-flight response causing heart palpitations, insomnia, inflammation, digestive imbalance, and headaches. The parasympathetic stimulating properties of some oils, especially ether compounds in oils such as anise seed and fennel, helps to reduce the fight-or-flight response and the conditions this causes.

STEP FOUR

Select a list of essential oils with compounds from the Structure-Effect groups indicated

The reference chart in Appendix C will aid in choosing essential oils for each molecular group. The following is a list of oils suitable for this example. Since each essential oil can have many common names, botanical identification is included for clarity:

- Balsam Poplar *(Poulus balsamifera)* – Sesquiterpene alcohols, sesquiterpene hydrocarbons
- Cape Chamomile *(Eriocephalus punctulatus)* – Esters, sesquiterpene hydrocarbons, sesquiterpene alcohols
- Cedarwood, Virginia *(Juniperus virginiana)* – Sesquiterpene hydrocarbons, sesquiterpene alcohols (cedrol)
- Clary Sage *(Salvia sclarea)* – Esters
- Copaiba *(Copaifera officinalis)* – Sesquiterpene hydrocarbons
- Eucalyptus globulus/Eucalyptus radiata – Oxides
- Fennel *(Foeniculum vulgare)* – Ethers
- Geranium *(Pelargonium x asperum)* – Esters, aldehydes, complex mixture of compounds
- German chamomile *(Matricaria recutita)* – Sesquiterpene alcohols, sesquiterpene hydrocarbons
- Ginger *(Zingiber officinale)* – Sesquiterpene alcohols, sesquiterpene hydrocarbons
- Helichrysum italicum – Sesquiterpene hydrocarbons, esters
- Lavender *(Lavandula angustifolia)* – Esters, complex mixture of compounds
- Lemongrass *(Cymbopogon flexuosus)* – Aldehydes
- Litsea cubeba – Aldehydes
- Peppermint *(Mentha piperita)* – Menthol (monoterpene alcohol)
- Spikenard *(Nardastachus jatamansi)* – Sesquiterpene alcohols, sesquiterpene hydrocarbons, aldehydes

- Tanacetum annuum – Sesquiterpene hydrocarbons
- Vetiver *(Vetiveria zizanoides)* – Sesquiterpene alcohols, sesquiterpene hydrocarbons
- Wintergreen *(Gaultheria procumbens)* – Esters (high in the analgesic ester, methyl salicylate)
- Ylang Ylang *(Cananga odorata)* – Esters, sesquiterpene hydrocarbons, sesquiterpene alcohols

There can be several essential oils to choose from. A sufficient list may be anywhere from 10–30 essential oils. It's not important how many essential oils are included on this list. What's important is to have a list that adequately provides the chemical families and properties needed with a variety of fragrance options (e.g., citrus, herbal, floral, earthy). It's always good to include essential oils with a complex structure, such as lavender, geranium and rose (the queen of complex structure).

STEP FIVE

Choose an application method

Before selecting the final essential oil(s) for the formula, consideration should be given to the method of application. The essential oils selected for an inhalation may differ quite a bit from a formula for a bath or topical application. The application is chosen based on what delivery will be best to achieve your therapeutic goal. In this example the chosen application is topical.

STEP SIX

Choose the essential oil(s) for the formula

Five essential oils will be selected for this example formula. A common essential oil formula can consist of 2–7 oils. The final selection of essential oils is based on many factors, including the experience of the formulator. Blending and formulating essential oils is an artful and scientific process. An abundance of blending instruction and techniques are available, offering tips on blending, formulating, formula percentages, application, and fragrance profiles of essential oils.

In this example, the five essential oils chosen from the list are:

- Lavender
- Cape chamomile
- Peppermint
- Fennel
- Vetiver

STEP SEVEN

Create a formula from the oils selected

This step defines the amount of each oil to be used. Your formula may be written out in percentage of each oil, in number of drops per oil, or in weight or volume amount of each oil. The amount of essential oil used in a formula is determined by the strength desired for treatment and can be from 0.05% to 50%. The most common percentage used is 2–3%. For skin care the common use is 1.5%.

The following formula is based on a 2.5% dilution of essential oil in a 30 ml bottle with a carrier oil. The total amount of essential oils needed is 15 drops.

Here's the formula for headache and the symptoms that were addressed:

Lavender – 3 drops

Cape chamomile – 2 drops

Peppermint – 5 drops

Fennel – 3 drops

Vetiver – 2 drops

The formula adds up to 15 drops of essential oil for a 2.5% dilution in a 30 mL container. (Refer to Chapter 8 of Jimm Harrison's *Aromatherapy* for more on formulating and blending essential oils.)

Repeat these steps with any therapeutic goal to determine which compound groups and which essential oils will best treat the conditions and symptoms of a goal.

Choosing Essential Oils Based On Properties, Not Chemistry

There are some essential oils that can be chosen due to being well known for the properties desired, even though this action does not appear apparent from the compound structure. If you focus only on the chemical structure of an oil as presented in the Structure-Effect Diagram you may miss using oils with desired therapeutic properties. The synergy of the chemical structure of the oil, along with unknown biological traits of the plant, create an identity and a use that cannot be determined by knowledge of the individual compounds.

Marjoram is an essential oil with well documented, and very effective, sedative properties and an ability to raise parasympathetic function (rest and digest mode). The chemistry that would support this is missing if looking for it on the structure diagram. It does have minor amounts of linalool and linalyl acetate, but not enough to explain its sedative potency. If marjoram were an unfamiliar oil and an attempt was made to understand it by looking at a GC to create a diagram, as shown in this book, anticipating its strong relaxing effects may be missed.

Frankincense is another example of chemistry that doesn't fully support properties associated with the essential oil. Its use to heal wounds is not found through observing the structure of the oil. Generally there are ketones or sesquiterpene alcohols in known skin regenerative essential oils. Frankincense may have 1–2% of either one. The anti-inflammatory action of frankincense isn't really supported by an obvious content of sesquiterpene compounds. Frankincense is an oil that has somehow gained a long list of claims, many that cannot be substantiated in any way. Be aware that many of the suggested uses of frankincense are related to known properties of boswellic acid, a compound found in the resin, not in the oil.

The interesting properties related to frankincense and the sedative properties of marjoram give credence to unknown factors that contribute to all essential oil therapeutic properties.

Conclusion

The easy formulating steps and the Structure-Effect Diagram are an efficient system for developing effective essential oil therapeutic recipes. The stronger your comprehension of the guidelines presented in this book, the greater your scientific, artistic, intuitive and logical approach will be. Remember, there is much more to essential oils than this limited reductionistic diagram. From here, your essential oil journey will integrate with an experience that lies deeper than and beyond individual chemical compounds. Essential oil chemistry and the Structure-Effect Diagram are one piece of an intricate puzzle that, when brought together, provides a mastery of boundless potential in stimulation of health, healing and planetary wellness with essential oils.

The chemistry is a foundational piece of a larger essential oil experience. You are the driver on this journey. It's your personal voyage. Your focus, the aspects of essential oils that stimulate your zeal, is of your choosing.

The path is one of passion, heart and intellectual stimulation, along with an obsessive undercurrent. It begins, and many are already here, with a personal and/or professional experience that demonstrates the awe-inspiring healing potential of essential oils and the ability to change lives for the better. These oils are dynamic, impressive and stunning. Embrace every facet of essential oils, from reductionistic science to complex biology and energetic vitality to spiritual resonance. Each piece cannot be isolated from the other. Essential oils are defined by synergy and are the most holistic heath-giving tool in the cabinet.

The fuel that powers the journey is information, education and, beyond all else, experience and developing a relationship with the oils. A study of the oils' interaction with olfaction, autonomic and central nervous systems, the biology and qi of the plants, cultivation and distillation, and the creative art and skill of blending and fragrancing with essential oils all come together in an integration that enables mastery over the healing inherent in the practice of aromatherapy.

In this book we've tapped into a useful tool to understand and explore one facet, the chemistry of the essential oils.

Let the journey continue.

*The Essential Oil and Aromatherapy Certificate Program at Bastyr University CCCE **www.bastyr.edu** provides a foundation of essential oil education.*

*Visit my website **www.jimmharrison.com** to experience all aspects of essential oil education, authentic essential oils, artistic blends and specialty carrier oils.*

Symptoms and the Structure-Effect Diagram

In the following diagrams there are symptoms or conditions listed in the areas of the compound groups that may be best for that particular symptom.

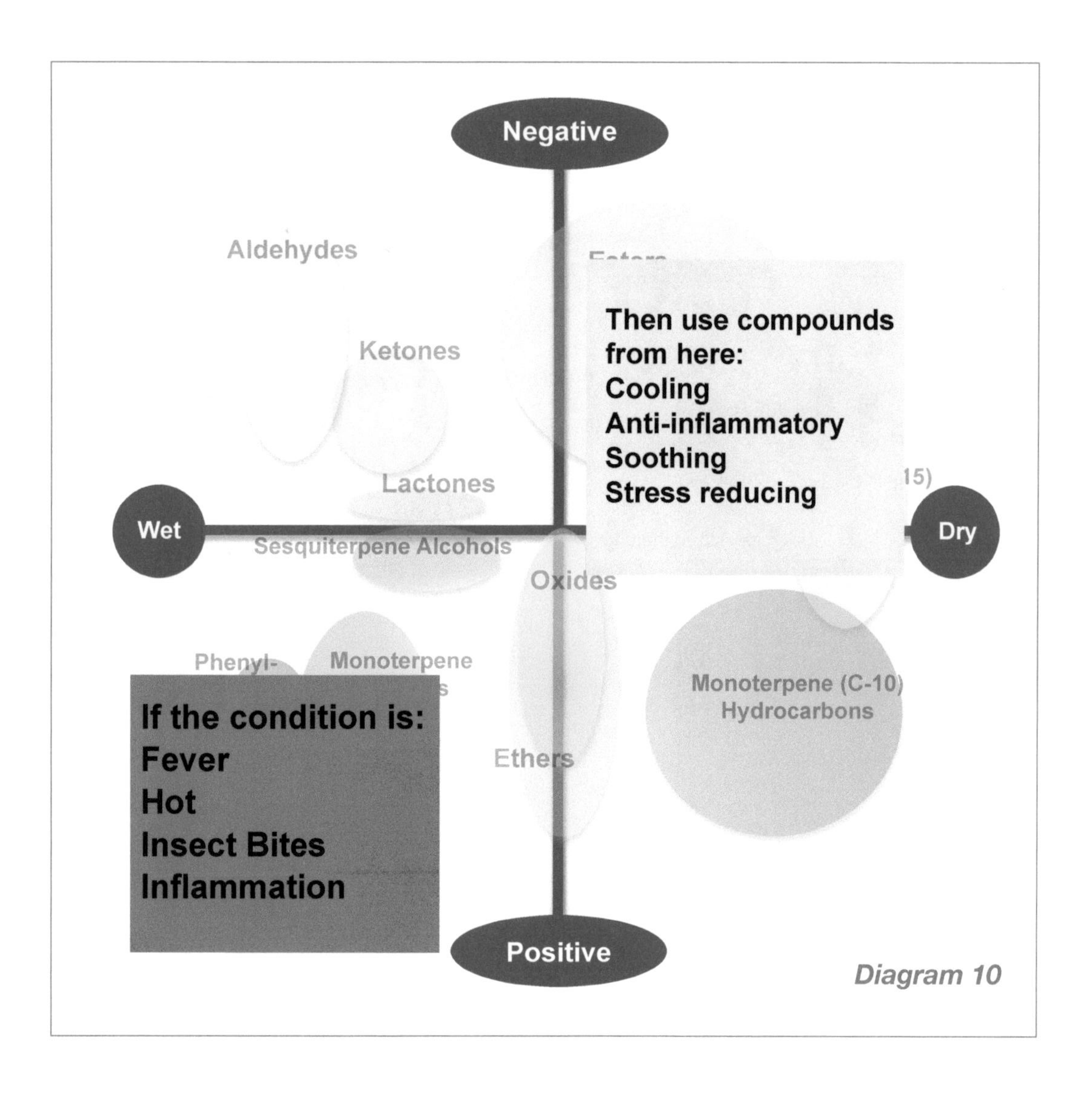

Diagram 10

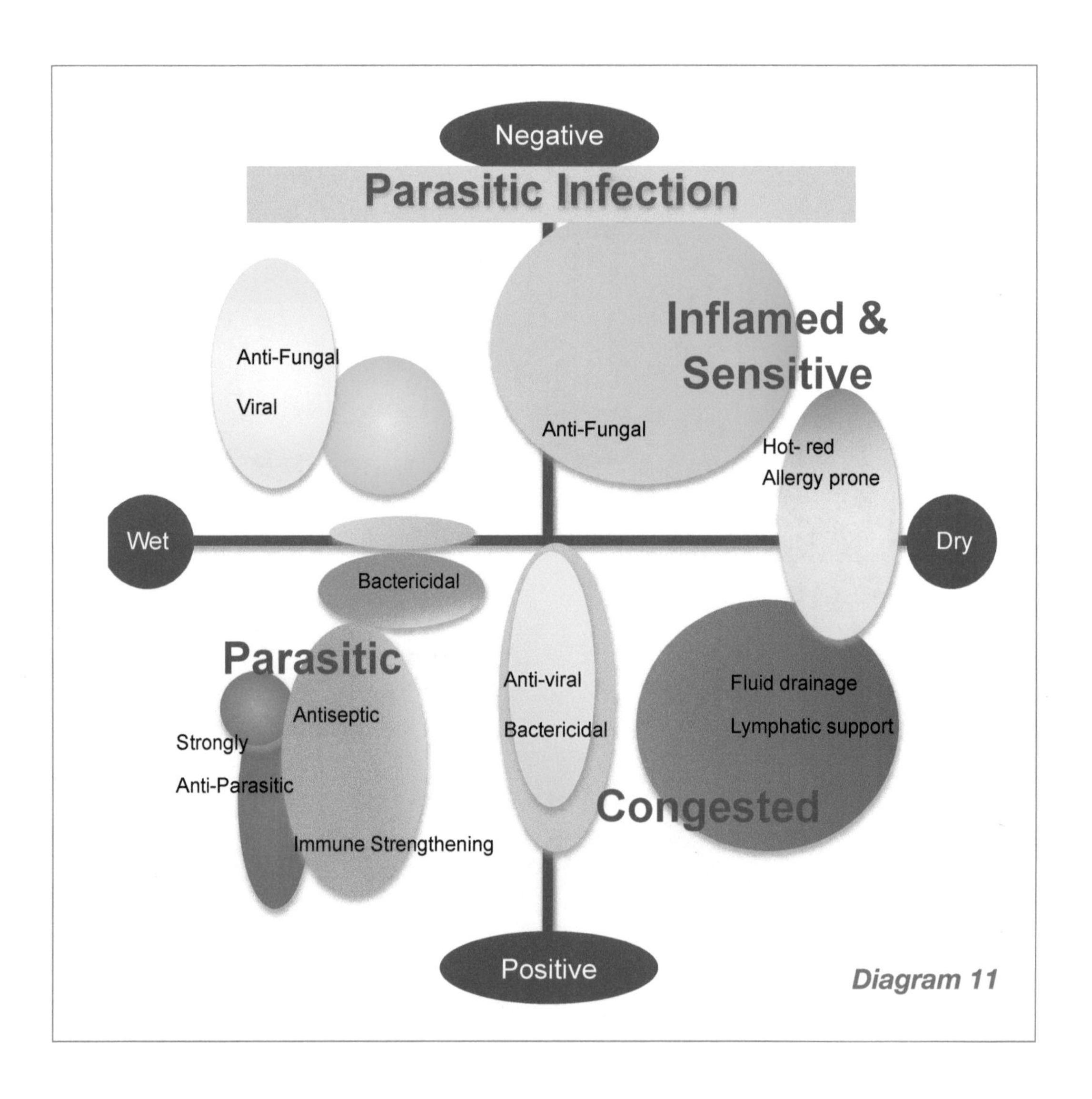
Negative
Parasitic Infection
Inflamed &
Sensitive
Anti-Fungal
Viral
Anti-Fungal
Hot- red
Allergy prone
Wet
Dry
Bactericidal
Parasitic
Antiseptic
Strongly
Anti-Parasitic
Immune Strengthening
Anti-viral
Bactericidal
Fluid drainage
Lymphatic support
Congested
Positive
Diagram 11

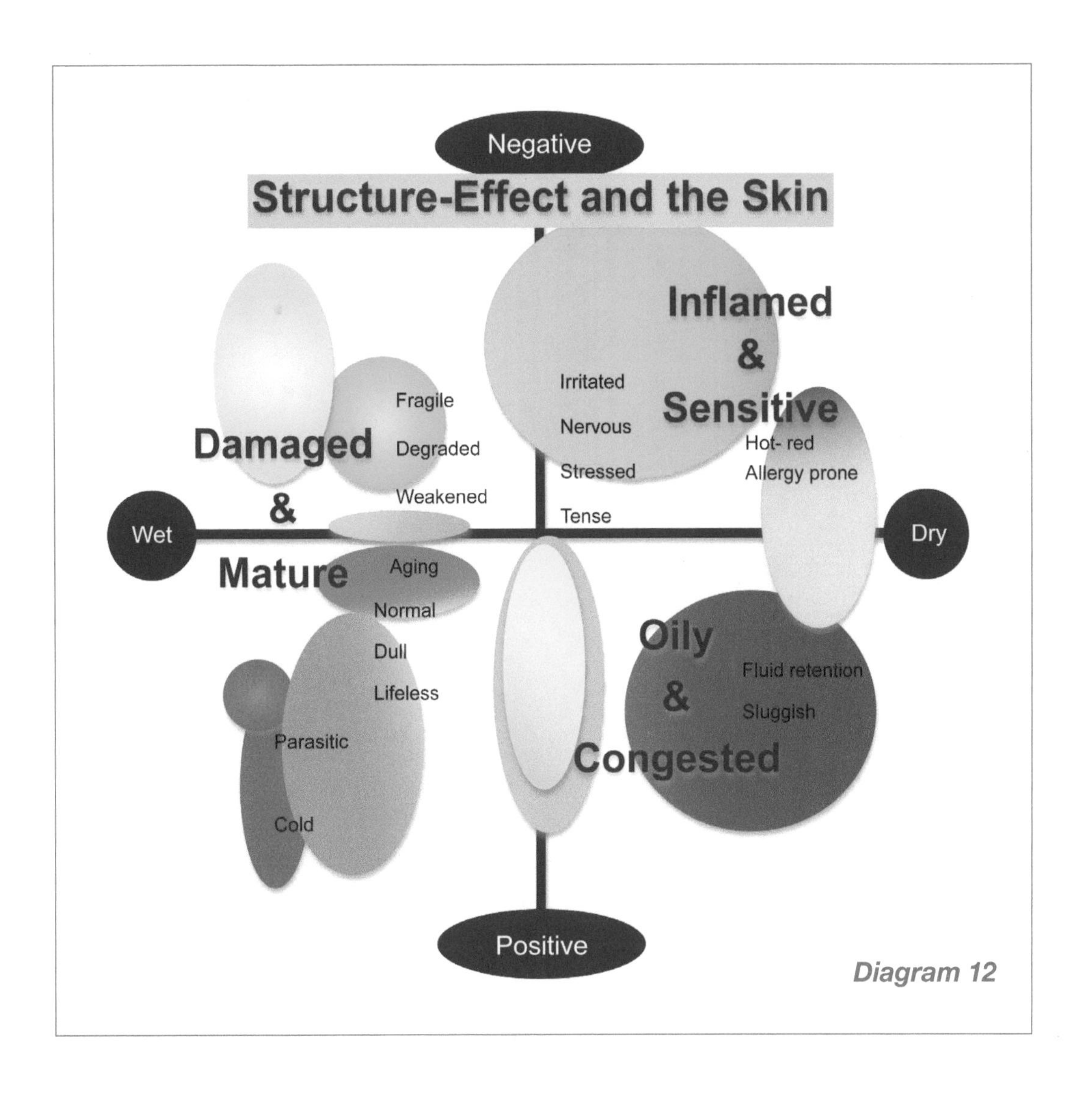
Negative
Structure-Effect and the Skin
Inflamed
&
Sensitive
Irritated
Nervous
Stressed
Tense
Hot- red
Allergy prone
Fragile
Degraded
Weakened
Damaged
&
Mature
Wet
Dry
Aging
Normal
Dull
Lifeless
Parasitic
Cold
Oily
&
Congested
Fluid retention
Sluggish
Positive
Diagram 12

APPENDIX B

Building a Structure-Effect Diagram: Rosemary CT Camphor

Rosemary *(Rosmarinus officinalis)* **chemotype (CT) Camphor**

Location: France and Spain

The plant *Rosmarinus officinalis* produces rosemary essential oils with a variety of chemotypes (CT), meaning that its main structure and compounds can vary greatly depending on geographical growing region, altitude and other conditions. In most cases, rosemary essential oil is sold and bought with a defined CT. When rosemary essential oil is sold without a defined CT, it is most likely a CT mix of 1,8-cineole and borneol. Other chemotypes of rosemary are: CT 1,8-cineole, CT verbenone, CT borneol and CT bornyl acetate. The oil in this sample is CT camphor, a rosemary with a higher content of the compound camphor. Rosemary CT camphor generally has a significant 1,8-cineole content and may also be called rosemary CT camphor/cineole.

Compound	Percentage	Compound Group
Monoterpene Hydrocarbons	alpha-pinene 4.4–22%, gamma-terpinene 0.5–7.8% Camphene 2.8–10%, (+)-Lim-onene 0–5.8%, beta-pinene 03–5%, beta-Myrcene 0–5.8%, 0–3.2%, para-Cymene 05–2.4%, Sabinene 0.1–1%	35–41%
Ketones	Camphor 17–27.3%, Verbenone 0–6.3%	17–28%
Oxide	1,8-Cineole 17–22.5%	17–22.5%
Monoterpene Alcohol	Borneol 2–9%, alpha-Terpineol 0–3.8%, Terpinen-4-ol 00.6–1.7%, linalool 0.9–1.5%	1–8%
Sesquiterpene Hydrocarbons	beta-Caryophyllene 0–2.5%,	0–2.5%
Esters	bornyl acetate 1–1.5%	1–1.5%
Ethers	methyleuganol 0–0.02%	0–0.02%

Rosemary CT Camphor Structure-Effect Diagram

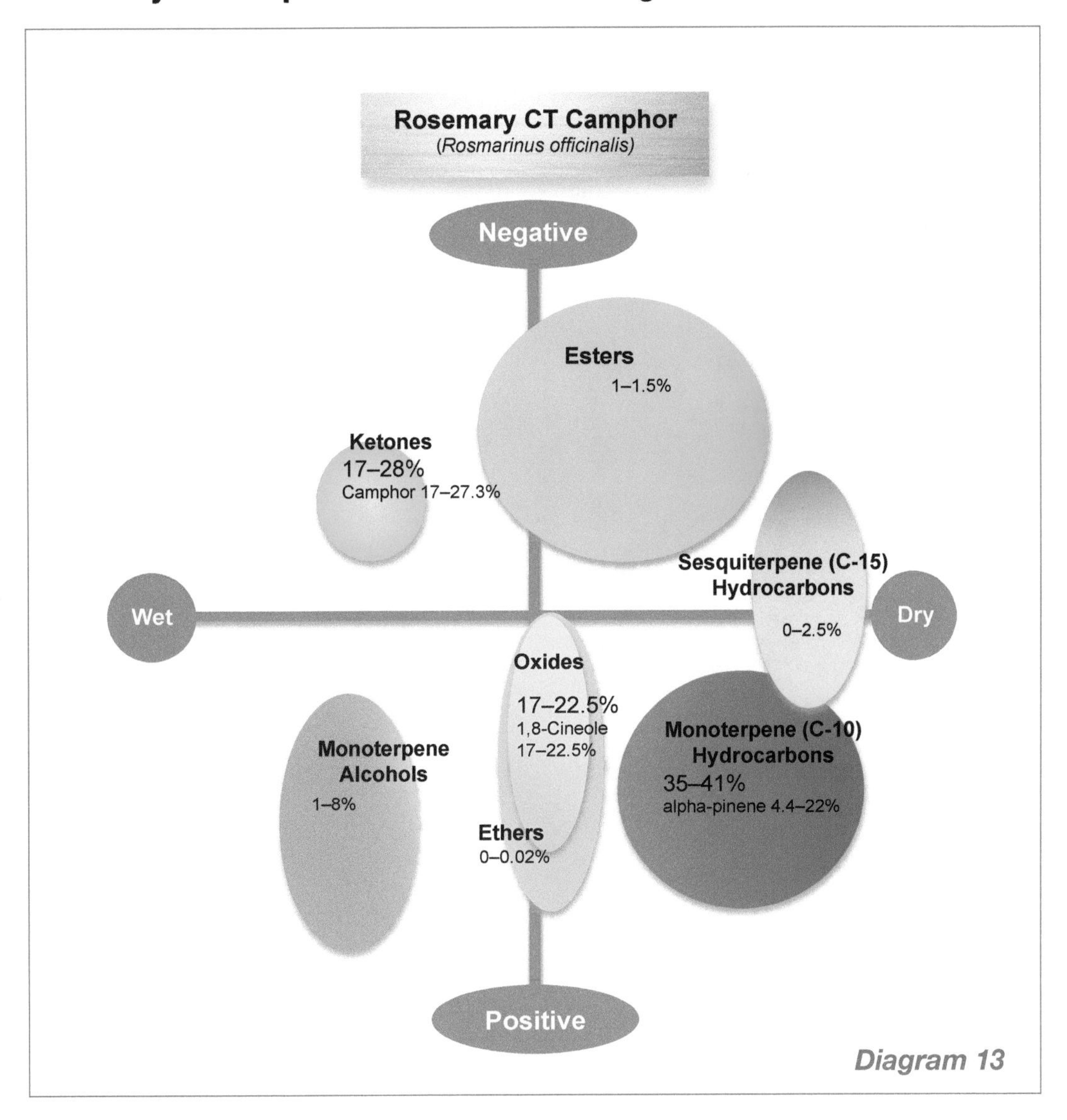

Diagram 13

APPENDIX C: CHEMICAL COMPOUNDS

C10 Hydrocarbons

Essential Oils	Total Range Monoterpene Hydrocarbons
Anise Seed *(Pimpinella anisum)*	tr–5%
Balsam Poplar *(Populus balsamifera)*	1.40%
Basil, Sweet, CT Linalool *(Ocymum basilicum)*	4–9%
Bay Laurel *(Laurus nobilis)*	19–25%
Bergamot *(Citrus aurantium bergamia)*	35–65%
Blue Tansy *(Tanacetum annuum)*	48–52.2%
Cardamon *(Elettaria cardamomum)*	4–14%
Carrot Seed *(Daucus carota)*	19–24%
Cedarwood, Atlas *(Cedrus atlantica)*	0.1–1%
Cedarwood, Virginia *(Juniperus virginiana)*	0.1–0.5%
Chamomile, Cape *(Eriocephalus punctulatus)*	7–8%

Essential Oils	Total Range Monoterpene Hydrocarbons
Chamomile, German *(Matricaria recutita)*	tr–1.3%
Chamomile, Roman *(Chamaemelum nobile)*	14.6–22%
Cinnamon, bark *(Cinnamomum zeylanicum)*	0.2–5%
Cistus *(Cistus ladaniferus)*	36–56%
Clary Sage *(Salvia sclarea)*	5–12%
Clove Bud *(Eugenia caryophyllata)*	tr–0.2%
Copaiba *(Copaifera officianalis)*	-
Coriander *(Coriandrum sativum)*	30–44.7%
Cypress *(Cupressus sempervirens)*	86.6–90.2%
Elemi *(Canarium luzonicum)*	70–77%
Eucalyptus citriodora	0–2.8%
Eucalyptus dives	28–44%

Essential Oils	Total Range Monoterpene Hydrocarbons
Eucalyptus globulus (not rectified)	20–25%
Eucalyptus radiata	5.4–12.2%
Eucalyptus staigeriana	19.9–75.1%
Fennel *(Foeniculum vulgaris)*	0.03–21%
Fir, Douglas *(Pseudotsuga menziesii)*	66–83%
Frankincense *(Boswellia carterii)*	77.8–89.5%
Galbanum *(Ferula galbaniflua)*	58.4–71%
Geranium *(Pelargonium x asperum)*	2–3%
Ginger *(Zinziber officinale)*	8.3–23%
Grapefruit *(Citrus paradisi)*	86.3–96%
Helichrysum *(Helichrysum italicum)*	5–10%
Inula graveolens	4.5–14.3%
Hyssop *(Hyssopus officinalis)*	2–18%

Essential Oils	Total Range Monoterpene Hydrocarbons
Jasmine Absolute *(Jasminum grandiflorum)*	0–1.1%
Juniper *(Juniperus communis)*	56.2–90.3%
Kunzea ambigua	34.6–52%
Lavandin *(Lavandula hybrida)*	1–6.6%
Lavender *(Lavandula angustifolia)*	0–4.3%
Lemon *(Citrus limon)*	83.4–97%
Lemongrass *(Cymbopogon citratus)*	0.34–23.9%
Lime *(Citrus limetta)*	78.3–89%
Litsea cubeba	11–29%
Manadrin *(Citrus reticulata)* peel	86–98%
Mandarin Petitgrain *(Citrus reticulata)* leaves	40–49%
Marjoram *(Origanum hortensis)*	13.6–31%
Melissa *(Melissa officinalis)*	3–6.1%

Essential Oils	Total Range Monoterpene Hydrocarbons
MQV/Niaouli *(Melaleuca quinquenervia viridiflora)* CT viridiflorol	tr–1%
Myrrh *(Commiphora myrrha)*	tr
Myrtle, Green *(Myrtus communis)*	23.8–73%
Neroli *(Citrus aurantium)* blossom	12.7–16%
Nutmeg *(Myristica fragrans)*	47–75.8%
Orange, Bitter *(Citrus aurantium)* peel	91.3–97.1%
Orange, Sweet *(Citrus sinensis)*	89–96%
Oregano *(Origanum vulgare)*	10–24%
Palmarosa *(Cymbopogon martinii)*	1.6–6%
Patchouli *(Pogostemon cablin)*	0–0.35%
Pepper, Black *(Piper nigrum)*	47–57%
Peppermint *(Mentha piperita)*	3–5%

Essential Oils	Total Range Monoterpene Hydrocarbons
Petitgrain *(Citrus aurantium)* leaves	6–10%
Pine *(Pinus sylvestris)*	27–96%
Plai *(Zingiber cassumunar)*	70.7–85%
Ravensare *(Ravensara aromatica)*	44.3–73.2%
Ravintsara/Ho Leaf CT Cineole *(Cinnamomum camphora)*	24–27%
Rose *(Rosa damascena)*	0.1–2.2%
Rosemary, CT Cineole *(Rosmarinus officinalis)*	16.7–28.5%
Rosemary, CT Verbenone *(Rosmarinus officinalis)*	7.8–63%
Sage *(Salvia officinalis)*	11–19%
Saint John's Wort *(Hypericum perforatum)*	18–45% Contains 38% alkane C9 hydrocarbons not grouped on the diagram

Essential Oils	Total Range Monoterpene Hydrocarbons
Sandalwood, Australian *(Santalum spicatum)*	-
Spearmint *(Mentha spicata)*	18–23%
Spike Lavender *(Lavandula latifolia)*	1.5–7.8%
Spikenard *(Nardostachys jatamansi)*	2–3.9%
Spruce, Black *(Picea mariana)*	38–50%
Tarragon *(Artemisia dracunculus)*	6–12%
Tea Tree *(Melaleuca alternifolia)*	37–48%
Thyme, CT Geraniol *(Thymus vulgaris)*	5–6.6%
Thyme, CT Linalool *(Thymus vulgaris)*	1.8–4.5%
Thyme, CT Thymol *(Thymus vulgaris)*	26–52.5%
Thymus satureioides	10–13%

Essential Oils	Total Range Monoterpene Hydrocarbons
Vetiver *(Vetiveria zizanoides)*	-
Vitex *(Vitex agnus castus)*	11–18%
Yarrow CT Chamazulene *(Achillea millefolium)*	10–16%
Ylang Ylang, Complete *(Cananga odorata)*	tr–2%

C15 Hydrocarbons

Essential Oils	Total Range Sesquiterpene Hydrocarbons
Anise Seed *(Pimpinella anisum)*	tr–2%
Balsam Poplar *(Populus balsamifera)*	37.78
Basil, Sweet, CT Linalool *(Ocimum basilicum)*	5–12%
Bay Laurel *(Laurus nobilis)*	tr–1.5%
Bergamot *(Citrus aurantium bergamia)*	tr–0.9%
Blue Tansy *(Tanacetum annuum)*	29–42%
Cardamon *(Elettaria cardamomum)*	tr
Carrot Seed *(Daucus carota)*	16–21%
Cedarwood, Atlas *(Cedrus atlantica)*	60–70%
Cedarwood, Virginia *(Juniperus virginiana)*	60–70%
Chamomile, Cape *(Eriocephalus punctulatus)*	4–5%
Chamomile, German *(Matricaria recutita)*	25–62%

Essential Oils	Total Range Sesquiterpene Hydrocarbons
Chamomile, Roman *(Chamaemelum nobile)*	tr–4%
Cinnamon, bark *(Cinnamomum zeylanicum)*	3–5%
Cistus *(Cistus ladaniferus)*	2.5–6%
Clary Sage *(Salvia sclarea)*	0.7–6.6%
Clove Bud *(Eugenia caryophyllata)*	0.6–14.3%
Copaiba *(Copaifera officianalis)*	80–92.7%
Coriander *(Coriandrum sativum)*	-
Cypress *(Cupressus sempervirens)*	0–1.84%
Elemi *(Canarium luzonicum)*	0–.42%
Eucalyptus citriodora	1.5–3%
Eucalyptus dives	0.1–1.6%
Eucalyptus globulus (not rectified)	0.8–3.5%
Eucalyptus radiata	0–1.6%

Essential Oils	Total Range Sesquiterpene Hydrocarbons
Eucalyptus staigeriana	-
Fennel *(Foeniculum vulgare)*	-
Fir, Douglas *(Pseudotsuga menziesii)*	0–9.4%
Frankincense *(Boswellia carterii)*	2–4%
Galbanum *(Ferula galbaniflua)*	20–26%
Geranium *(Pelargonium x asperum)*	4–17.3%
Ginger *(Zinziber officinale)*	52–82%
Grapefruit *(Citrus paradisi)*	0–2.2%
Helichrysum *(Helichrysum italicum)*	20.4–36%
Inula graveolens	5.04–8.4%
Hyssop *(Hyssopus officinalis)*	-
Jasmine Absolute *(Jasminum grandiflorum)*	0–6%
Juniper *(Juniperus communis)*	3.1–14%

Essential Oils	Total Range Sesquiterpene Hydrocarbons
Kunzea ambigua	4.8–7.2%
Lavandin *(Lavandula hybrida)*	0.4–2%
Lavender *(Lavandula angustifolia)*	1–1.8%
Lemon *(Citrus limon)*	0–0.9%
Lemongrass *(Cymbopogon citratus)*	0.2–3%
Lime *(Citrus limetta)*	1.6–5%
Litsea cubeba	-
Manadrin *(Citrus reticulata)* peel	0–0.4%
Mandarin Petitgrain *(Citrus reticulata)* leaves	1.2–1.5%
Marjoram *(Origanum hortensis)*	0–3.3%
Melissa *(Melissa officinalis)*	6–49%
MQV *(Melaleuca quinquenervia viridiflora)* CT viridiflorol	43–50%
Myrrh *(Commiphora myrrha)*	23.4%

Essential Oils	Total Range Sesquiterpene Hydrocarbons
Myrtle, Green *(Myrtus communis)*	0–0.6%
Neroli *(Citrus aurantium)* blossom	0–0.7%
Nutmeg *(Myristica fragrans)*	0–1%
Orange, Bitter *(Citrus aurantium)* peel	-
Orange, Sweet *(Citrus sinensis)*	0–1.5%
Oregano *(Origanum vulgaris)*	1.4–3.1%
Palmarosa *(Cymbopogon martini)*	0.9–2.6%
Patchouli *(Pogostemon cablin)*	48–56%
Pepper, Black *(Piper nigrum)*	9.8–52.8%
Peppermint *(Mentha piperita)*	1–6%
Petitgrain *(Citrus aurantium)* leaves	-
Plai *(Zingiber cassumunar)*	1.8–6%
Pine *(Pinus sylvestris)*	2–39%

Essential Oils	Total Range Sesquiterpene Hydrocarbons
Ravensare *(Ravensara aromatica)*	2.6–19%
Ravintsara/Ho Leaf CT Cineole *(Cinnamomum camphora)*	0.6–3.1%
Rose *(Rosa damascena)*	1–4% Contains 19–24% C19 alkane and alkene hydrocarbons not present on the diagram
Rosemary, CT Cineole *(Rosmarinus officinalis)*	3.6–10.8%
Rosemary, CT Verbenone *(Rosmarinus officinalis)*	0.7–3.3%
Sage *(Salvia officinalis)*	0.2–9.7%
Saint John's Wort *(Hypericum perforatum)*	1.5–4%
Sandalwood, Australian *(Santalum spicatum)*	1–3%
Spearmint *(Mentha viridis)*	0–2%

Essential Oils	Total Range Sesquiterpene Hydrocarbons
Spike Lavender *(Lavandula latifolia)*	1.3–5.8%
Spikenard *(Nardostachys jatamansi)*	30–45.4%
Spruce, Black *(Picea mariana)*	tr–2.5%
Tarragon *(Artemisia dracunculus)*	tr–3%
Tea Tree *(Melaleuca alternifolia)*	5–8%
Thyme, CT Geraniol *(Thymus vulgaris)*	0–2%
Thyme, CT Linalool *(Thymus vulgaris)*	0–2%
Thyme, CT Thymol *(Thymus vulgaris)*	1.3–3.1%
Thymus satureioides	2.3–4%
Vetiver *(Vetiveria zizanoides)*	28–35%
Vitex *(Vitex agnus castus)*	36–45%
Yarrow CT Chamazulene *(Achillea millefolium)*	30–52%
Ylang Ylang, Complete *(Cananga odorata)*	37–56.3%

C10 Alcohols

Essential Oils	Total Range Monoterpene Alcohols
Anise Seed *(Pimpinella anisum)*	0.1–3.5%
Balsam Poplar *(Populus balsamifera)*	0.18
Basil, Sweet, CT Linalool *(Ocimum basilicum)*	34–59%
Bay Laurel *(Laurus nobilis)*	6.5–10%
Bergamot *(Citrus aurantium bergamia)*	1.7–20%
Blue Tansy *(Tanacetum annuum)*	1.4–7.2%
Cardamon *(Elettaria cardamomum)*	1–14%
Carrot Seed *(Daucus carota)*	0–2.2%
Cedarwood, Atlas *(Cedrus atlantica)*	tr
Cedarwood, Virginia *(Juniperus virginiana)*	tr
Chamomile, Cape *(Eriocephalus punctulatus)*	4–5.5%
Chamomile, German *(Matricaria recutita)*	tr–1%

Essential Oils	Total Range Monoterpene Alcohols
Chamomile, Roman *(Chamaemelum nobile)*	3–12%
Cinnamon, bark *(Cinnamomum zeylanicum)*	0.7–6%
Cistus *(Cistus ladaniferus)*	9–19.8%
Clary Sage *(Salvia sclarea)*	9.6–20.5%
Clove Bud *(Eugenia caryophyllata)*	tr
Copaiba *(Copaifera officianalis)*	tr–5%
Coriander *(Coriandrum sativum)*	60–87%
Cypress *(Cupressus sempervirens)*	tr–5.8%
Elemi *(Canarium luzonicum)*	0–3%
Eucalyptus citriodora	3.99–24.8%
Eucalyptus dives	6–8.4%
Eucalyptus globulus (not rectified)	2–5%
Eucalyptus radiata	8–34.8%

Essential Oils	Total Range Monoterpene Alcohols
Eucalyptus staigeriana	1–7.1%
Fennel *(Foeniculum vulgare)*	-
Fir, Douglas *(Pseudotsuga menziesii)*	2–6.6%
Frankincense *(Boswellia carterii)*	1.8–3.2%
Galbanum *(Ferula galbaniflua)*	0.4–4.6%
Geranium *(Pelargonium x asperum)*	39–73%
Ginger *(Zinziber officinale)*	0.7–3%
Grapefruit *(Citrus paradisi)*	0–0.2%
Helichrysum *(Helichrysum italicum)*	-
Inula graveolens	14.8–17.2%
Hyssop *(Hyssopus officinalis)*	0.4–2%
Jasmine Absolute *(Jasminum grandiflorum)*	8.5–11%
Juniper *(Juniperus communis)*	2–19%

Essential Oils	Total Range Monoterpene Alcohols
Kunzea ambigua	4–6.8%
Lavandin (Lavandula hybrida)	22–42%
Lavender *(Lavandula angustifolia)*	30–47%
Lemon *(Citrus limon)*	0.1–9%
Lemongrass *(Cymbopogon citratus)*	4.5–10%
Lime *(Citrus limetta)*	2.5–11.6%
Litsea cubeba	2–3.3%
Manadrin *(Citrus reticulata)* peel	-
Mandarin Petitgrain *(Citrus reticulata)* leaves	0–1.2%
Marjoram *(Origanum hortensis)*	19.6–25%
Melissa *(Melissa officinalis)*	1.7–9.2%
MQV *(Melaleuca quinquenervia viridiflora)* CT viridiflorol	tr
Myrrh *(Commiphora myrrha)*	-
Myrtle, Green *(Myrtus communis)*	1.7–6%

Essential Oils	Total Range Monoterpene Alcohols
Neroli *(Citrus aurantium)* blossom	36.2–66.3%
Nutmeg *(Myristica fragrans)*	5–8%
Orange, Bitter *(Citrus aurantium)* peel	0.1–2.0%
Orange, Sweet *(Citrus sinensis)*	0–5%
Oregano *(Origanum vulgaris)*	0.6–2.1%
Palmarosa *(Cymbopogon martini)*	77.1–85.5%
Patchouli *(Pogostemon cablin)*	-
Pepper, Black *(Piper nigrum)*	tr–0.35%
Peppermint *(Mentha piperita)*	38–49%
Petitgrain *(Citrus aurantium)* leaves	29–32.6%
Pine *(Pinus sylvestris)*	0–2.7%
Plai *(Zingiber cassumunar)*	21.7–41.7%
Ravensare *(Ravensara aromatica)*	3–8%

Essential Oils	Total Range Monoterpene Alcohols
Ravintsara/Ho Leaf CT Cineole *(Cinnamomum camphora)*	16–28%
Rose *(Rosa damascena)*	36–50.7%
Rosemary, CT Cineole *(Rosmarinus officinalis)*	6–9%
Rosemary, CT Verbenone *(Rosmarinus officinalis)*	7–22%
Sage *(Salvia officinalis)*	1.5–24%
Saint John's Wort *(Hypericum perforatum)*	0–0.4%
Sandalwood, Australian *(Santalum spicatum)*	0–3.1%
Spearmint *(Mentha viridis)*	3–4.2%
Spike Lavender *(Lavandula latifolia)*	30.9–49.6%
Spikenard *(Nardostachys jatamansi)*	0–0.5%
Spruce, Black *(Picea mariana)*	1.4–2.1%

Essential Oils	Total Range Monoterpene Alcohols
Tarragon *(Artemisia dracunculus)*	1–5.6%
Tea Tree *(Melaleuca alternifolia)*	38.5–44%
Thyme, CT Geraniol *(Thymus vulgaris)*	30–34.4%
Thyme, CT Linalool *(Thymus vulgaris)*	49–53.3%
Thyme, CT Thymol *(Thymus vulgaris)*	0.6–3%
Thymus satureioides	31.5–35%
Vetiver *(Vetiveria zizanoides)*	-
Vitex *(Vitex agnus castus)*	7–13%
Yarrow CT Chamazulene *(Achillea millefolium)*	2–6.5%
Ylang Ylang, Complete *(Cananga odorata)*	8–12%

C15 Alcohols

Essential Oils	Total Range Sesquiterpene Alcohols
Anise Seed *(Pimpinella anisum)*	-
Balsam Poplar *(Populus balsamifera)*	56.7%
Basil, Sweet, CT Linalool *(Ocimum basilicum)*	0.5–5%
Bay Laurel *(Laurus nobilis)*	-
Bergamot *(Citrus aurantium bergamia)*	-
Blue Tansy *(Tanacetum annuum)*	0.63–6.7%
Cardamon *(Elettaria cardamomum)*	0.1–2.7%
Carrot Seed *(Daucus carota)*	36.1–74%
Cedarwood, Atlas *(Cedrus atlantica)*	8–11%
Cedarwood, Virginia *(Juniperus virginiana)*	23–24%
Chamomile, Cape *(Eriocephalus punctulatus)*	1–3%
Chamomile, German *(Matricaria recutita)*	3–8%

Essential Oils	Total Range Sesquiterpene Alcohols
Chamomile, Roman *(Chamaemelum nobile)*	-
Cinnamon, bark *(Cinnamomum zeylanicum)*	-
Cistus *(Cistus ladaniferus)*	3–5%
Clary Sage *(Salvia sclarea)*	0–1.33% with C20 alcohol, sclareol
Clove Bud *(Eugenia caryophyllata)*	-
Copaiba *(Copaifera officianalis)*	0–8.4%
Coriander, Seed *(Coriandrum sativum)*	-
Cypress *(Cupressus sempervirens)*	1.1–7.0%
Elemi *(Canarium luzonicum)*	2.8–17%
Eucalyptus citriodora	-
Eucalyptus dives	tr
Eucalyptus globulus (not rectified)	tr–5.3%

Essential Oils	Total Range Sesquiterpene Alcohols
Eucalyptus radiata	-
Eucalyptus staigeriana	-
Fennel *(Foeniculum vulgare)*	-
Fir, Douglas *(Pseudotsuga menziesii)*	-
Frankincense *(Boswellia carterii)*	0–15%
Galbanum *(Ferula galbaniflua)*	1–4%
Geranium *(Pelargonium x asperum)*	tr–8.9%
Ginger *(Zinziber officinale)*	0–2.5%
Grapefruit *(Citrus paradisi)*	-
Helichrysum *(Helichrysum italicum)*	-
Inula graveolens	2.35–10%
Hyssop *(Hyssopus officinalis)*	-
Jasmine Absolute *(Jasminum grandiflorum)*	5–8.8%
Juniper *(Juniperus communis)*	-

Essential Oils	Total Range Sesquiterpene Alcohols
Kunzea ambigua	11–24%
Lavandin *(Lavandula hybrida)*	-
Lavender *(Lavandula angustifolia)*	-
Lemon *(Citrus limon)*	tr
Lemongrass *(Cymbopogon citratus)*	tr
Lime *(Citrus limetta)*	tr
Litsea cubeba	-
Manadrin *(Citrus reticulata)* peel	-
Mandarin Petitgrain *(Citrus reticulata)*	-
Marjoram *(Origanum hortensis)*	-
Melissa *(Melissa officinalis)*	0–1.7%
MQV *(Melaleuca quinquenervia viridiflora)* CT viridiflorol	44–70%
Myrrh *(Commiphora myrrha)*	1.3–2%
Myrtle, Green *(Myrtus communis)*	tr

Essential Oils	Total Range Sesquiterpene Alcohols
Neroli *(Citrus aurantium)* blossom	2.0–5.6%
Nutmeg *(Myristica fragrans)*	-
Orange, Bitter *(Citrus aurantium)* peel	tr–1%
Orange, Sweet *(Citrus sinensis)*	-
Oregano *(Origanum vulgaris)*	-
Palmarosa *(Cymbopogon martini)*	0.8–7%
Patchouli *(Pogostemon cablin)*	29–35.7%
Pepper, Black *(Piper nigrum)*	0–1.5%
Peppermint *(Mentha piperita)*	0–0.6%
Petitgrain *(Citrus aurantium)* leaves	-
Pine *(Pinus sylvestris)*	0–2.8%
Plai *(Zingiber cassumunar)*	0–0.2%
Ravensare *(Ravensara aromatica)*	0–4.9%

Essential Oils	Total Range Sesquiterpene Alcohols
Ravintsara/Ho Leaf CT Cineole *(Cinnamomum camphora)*	0.5–3.2%
Rose *(Rosa damascena)*	0–2.7%
Rosemary, CT Cineole *(Rosmarinus officinalis)*	0–0.4%
Rosemary, CT Verbenone *(Rosmarinus officinalis)*	-
Sage *(Salvia officinalis)*	0.2–2%
Saint John's Wort *(Hypericum perforatum)*	0–3%
Sandalwood, Australian *(Santalum spicatum)*	68–72%
Spearmint *(Mentha viridis)*	1.5–2.6%
Spike Lavender *(Lavandula latifolia)*	-
Spikenard *(Nardostachys jatamansi)*	6–14.3%
Spruce, Black *(Picea mariana)*	0–0.3%

Essential Oils	Total Range Sesquiterpene Alcohols
Tarragon *(Artemisia dracunculus)*	-
Tea Tree *(Melaleuca alternifolia)*	tr–2.0%
Thyme, CT Geraniol *(Thymus vulgaris)*	0–3%
Thyme, CT Linalool *(Thymus vulgaris)*	-
Thyme, CT Thymol *(Thymus vulgaris)*	-
Thymus satureioides	-
Vetiver *(Vetiveria zizanoides)*	12–38%
Vitex *(Vitex agnus castus)*	9–11%
Yarrow CT Chamazulene *(Achillea millefolium)*	0–0.3%
Ylang Ylang, Complete *(Cananga odorata)*	3.3–9%

Phenols

Essential Oils	Total Range Phenols
Anise Seed *(Pimpinella anisum)*	-
Balsam Poplar *(Populus balsamifera)*	-
Basil, Sweet, CT Linalool *(Ocimum basilicum)*	-
Bay Laurel *(Laurus nobilis)*	0–0.7%
Bergamot *(Citrus aurantium bergamia)*	-
Blue Tansy *(Tanacetum annuum)*	0.52–1.8%
Cardamon *(Elettaria cardamomum)*	-
Carrot Seed *(Daucus carota)*	-
Cedarwood, Atlas *(Cedrus atlantica)*	-
Cedarwood, Virginia *(Juniperus virginiana)*	-
Chamomile, Cape *(Eriocephalus punctulatus)*	tr
Chamomile, German *(Matricaria recutita)*	-

Essential Oils	Total Range Phenols
Chamomile, Roman *(Chamaemelum nobile)*	-
Cinnamon, bark *(Cinnamomum zeylanicum)*	-
Cistus *(Cistus ladaniferus)*	tr–2.3%
Clary Sage *(Salvia sclarea)*	-
Clove Bud *(Eugenia caryophyllata)*	see Phenylpropanoids (Hot)
Copaiba *(Copaifera officianalis)*	-
Coriander *(Coriandrum sativum)*	-
Cypress *(Cupressus sempervirens)*	-
Elemi *(Canarium luzonicum)*	-
Eucalyptus citriodora	-
Eucalyptus dives	-
Eucalyptus globulus (not rectified)	tr
Eucalyptus radiata	-

Essential Oils	Total Range Phenols
Eucalyptus staigeriana	-
Fennel *(Foeniculum vulgare)*	-
Fir, Douglas *(Pseudotsuga menziesii)*	0–2.9%
Frankincense *(Boswellia carterii)*	-
Galbanum *(Ferula galbaniflua)*	-
Geranium *(Pelargonium x asperum)*	-
Ginger *(Zinziber officinale)*	-
Grapefruit *(Citrus paradisi)*	-
Helichrysum *(Helichrysum italicum)*	-
Inula graveolens	
Hyssop *(Hyssopus officinalis)*	-
Jasmine *(Jasminum grandiflorum)*	0–2.4%
Juniper *(Juniperus communis)*	-
Kunzea ambigua	-

Essential Oils	Total Range Phenols
Lavandin *(Lavandula hybrida)*	-
Lavender *(Lavandula angustifolia)*	-
Lemon *(Citrus limon)*	-
Lemongrass *(Cymbopogon citratus)*	-
Lime *(Citrus limetta)*	-
Litsea cubeba	-
Manadrin *(Citrus reticulata)* peel	-
Mandarin Petitgrain *(Citrus reticulata)* leaves	-
Marjoram *(Origanum hortensis)*	-
Melissa *(Melissa officinalis)*	tr–0.4%
MQV *(Melaleuca quinquenervia viridiflora)* CT viridiflorol	-
Myrrh *(Commiphora myrrha)*	-
Myrtle, Green *(Myrtus communis)*	

Essential Oils	Total Range Phenols
Neroli *(Citrus aurantium)* blossom	-
Nutmeg *(Myristica fragrans)*	-
Orange, Bitter *(Citrus aurantium)* peel	-
Orange, Sweet *(Citrus sinensis)*	-
Oregano *(Origanum vulgaris)*	67–86%
Palmarosa *(Cymbopogon martini)*	-
Patchouli *(Pogostemon cablin)*	-
Pepper, Black *(Piper nigrum)*	-
Peppermint *(Mentha piperita)*	-
Petitgrain *(Citrus aurantium)* leaves	-
Pine *(Pinus sylvestris)*	-
Plai *(Zingiber cassumunar)*	-
Ravensare *(Ravensara aromatica)*	-

Essential Oils	Total Range Phenols
Ravintsara/Ho Leaf CT Cineole *(Cinnamomum camphora)*	-
Rose *(Rosa damascena)*	-
Rosemary, CT Cineole *(Rosmarinus officinalis)*	tr
Rosemary, CT Verbenone *(Rosmarinus officinalis)*	-
Sage *(Salvia officinalis)*	-
Saint John's Wort *(Hypericum perforatum)*	-
Sandalwood, Australian *(Santalum spicatum)*	-
Spearmint *(Mentha viridis)*	-
Spike Lavender *(Lavandula latifolia)*	-
Spikenard *(Nardostachys jatamansi)*	-
Spruce, Black *(Picea mariana)*	-

Essential Oils	Total Range Phenols
Tarragon *(Artemisia dracunculus)*	-
Tea Tree *(Melaleuca alternifolia)*	-
Thyme, CT Geraniol *(Thymus vulgaris)*	-
Thyme, CT Linalool *(Thymus vulgaris)*	tr–1.2%
Thyme, CT Thymol *(Thymus vulgaris)*	29–56%
Thymus satureioides	26–30%
Vetiver *(Vetiveria zizanoides)*	-
Vitex *(Vitex agnus castus)*	0–1%
Yarrow CT Chamazulene *(Achillea millefolium)*	-
Ylang Ylang, Complete *(Cananga odorata)*	-

Esters

Essential Oils	Total Range Esters
Anise Seed *(Pimpinella anisum)*	0–2%
Balsam Poplar *(Populus balsamifera)*	-
Basil, Sweet, CT Linalool *(Ocimum basilicum)*	tr–4%
Bay Laurel *(Laurus nobilis)*	7–14%
Bergamot *(Citrus aurantium bergamia)*	17.8–41%
Blue Tansy *(Tanacetum annuum)*	tr–1%
Cardamon *(Elettaria cardamomum)*	30–47%
Carrot Seed *(Daucus carota)*	0.2–4%
Cedarwood, Atlas *(Cedrus atlantica)*	-
Cedarwood, Virginia *(Juniperus virginiana)*	-
Chamomile, Cape *(Eriocephalus punctulatus)*	70–80%
Chamomile, German *(Matricaria recutita)*	-

Essential Oils	Total Range Esters
Chamomile, Roman *(Chamaemelum nobile)*	75–85%
Cinnamon, bark *(Cinnamomum zeylanicum)*	1–19%
Cistus *(Cistus ladaniferus)*	0.9–4.8%
Clary Sage *(Salvia sclarea)*	0.67–74.8%
Clove Bud *(Eugenia caryophyllata)*	0.2–11.6%
Copaiba *(Copaifera officianalis)*	-
Coriander *(Coriandrum sativum)*	1–3.6%
Cypress *(Cupressus sempervirens)*	1.9–8.5%
Elemi *(Canarium luzonicum)*	-
Eucalyptus citriodora	1–10%
Eucalyptus dives	-
Eucalyptus globulus (not rectified)	tr
Eucalyptus radiata	0.6–3%
Eucalyptus staigeriana	0.9–6.5%

Essential Oils	Total Range Esters
Fennel *(Foeniculum vulgare)*	-
Fir, Douglas *(Pseudotsuga menziesii)*	10–24%
Frankincense *(Boswellia carterii)*	1.1–2.2%
Galbanum *(Ferula galbaniflua)*	-
Geranium *(Pelargonium x asperum)*	12.4–31.7%
Ginger *(Zinziber officinale)*	tr
Grapefruit *(Citrus paradisi)*	-
Helichrysum *(Helichrysum italicum)*	8–50%
Inula graveolens	46.1–48.2%
Hyssop *(Hyssopus officinalis)*	-
Jasmine, Absolute *(Jasminum grandiflorum)*	49–58%
Juniper *(Juniperus communis)*	-
Kunzea ambigua	0–0.5%
Lavandin *(Lavandula hybrida)*	38–41.2%

Essential Oils	Total Range Esters
Lavender *(Lavandula angustifolia)*	30–50%
Lemon *(Citrus limon)*	tr–1.5%
Lemongrass *(Cymbopogon citratus)*	0.4–4.3%
Lime *(Citrus limetta)*	tr–1%
Litsea cubeba	0–1.6%
Manadrin *(Citrus reticulata)* peel	0–0.4%
Mandarin Petitgrain *(Citrus reticulata)* leaves	43–52%
Marjoram *(Origanum hortensis)*	9–11%
Melissa *(Melissa officinalis)*	2.2–13%
MQV *(Melaleuca quinquenervia viridiflora)* CT viridiflorol	-
Myrrh *(Commiphora myrrha)*	-
Myrtle, Green *(Myrtus communis)*	4–27.7%
Neroli *(Citrus aurantium)* blossom	5–10.2%
Nutmeg *(Myristica fragrans)*	0–0.3%

Essential Oils	Total Range Esters
Orange, Bitter *(Citrus aurantium)* peel	tr–2%
Orange, Sweet *(Citrus sinensis)*	0–0.6%
Oregano *(Origanum vulgaris)*	-
Palmarosa *(Cymbopogon martini)*	0.5–10.7%
Patchouli *(Pogostemon cablin)*	-
Pepper, Black *(Piper nigrum)*	-
Peppermint *(Mentha piperita)*	2–10%
Peru Balsam *(Myroxylon pereira)*	59.4–91.7%
Petitgrain *(Citrus aurantium)* leaves	58–71%
Pine *(Pinus sylvestris)*	0–4.2%
Plai *(Zingiber cassumunar)*	0–0.3%
Ravensare *(Ravensara aromatica)*	tr–6%
Ravintsara/Ho Leaf CT Cineole *(Cinnamomum camphora)*	-

Essential Oils	Total Range Esters
Rose *(Rosa damascena)*	1–2.8%
Rosemary, CT Cineole *(Rosmarinus officinalis)*	0.3–1.2%
Rosemary, CT Verbenone *(Rosmarinus officinalis)*	2.9–11.3%
Sage *(Salvia officinalis)*	2.6–6%
Saint John's Wort *(Hypericum perforatum)*	0–1.5%
Sandalwood, Australian *(Santalum spicatum)*	-
Spearmint *(Mentha viridis)*	-
Spike Lavender *(Lavandula latifolia)*	0–0.5%
Spikenard *(Nardostachys jatamansi)*	-
Spruce, Black *(Picea mariana)*	24–37%
Tarragon *(Artemisia dracunculus)*	0–3%
Tea Tree *(Melaleuca alternifolia)*	-

Essential Oils	Total Range Esters
Thyme, CT Geraniol *(Thymus vulgaris)*	36–42%
Thyme, CT Linalool *(Thymus vulgaris)*	0–2%
Thyme, CT Thymol *(Thymus vulgaris)*	0–tr%
Thymus satureioides	1–2.5%
Vetiver *(Vetiveria zizanoides)*	-
Vitex *(Vitex agnus castus)*	0.3–7.8%
Yarrow CT Chamazulene *(Achillea millefolium)*	2–4.3%
Ylang Ylang, Complete *(Cananga odorata)*	21–29%

Aldehydes

Essential Oils	Total Range Aldehydes
Anise Seed *(Pimpinella anisum)*	0.1–1.4%
Balsam Poplar *(Populus balsamifera)*	-
Basil, Sweet, CT Linalool *(Ocimum basilicum)*	-
Bay Laurel *(Laurus nobilis)*	-
Bergamot *(Citrus aurantium bergamia)*	tr–0.3%
Blue Tansy *(Tanacetum annuum)*	-
Cardamon *(Elettaria cardamomum)*	-
Carrot Seed *(Daucus carota)*	-
Cedarwood, Atlas *(Cedrus atlantica)*	-
Cedarwood, Virginia *(Juniperus virginiana)*	-
Chamomile, Cape *(Eriocephalus punctulatus)*	-
Chamomile, German *(Matricaria recutita)*	tr

Essential Oils	Total Range Aldehydes
Chamomile, Roman *(Chamaemelum nobile)*	0–1%
Cinnamon, bark *(Cinnamomum zeylanicum)*	see phenylpropanoids (hot)
Cistus *(Cistus ladaniferus)*	tr–2.6%
Clary Sage *(Salvia sclarea)*	-
Clove Bud *(Eugenia caryophyllata)*	-
Copaiba *(Copaifera officianalis)*	-
Coriander *(Coriandrum sativum)*	-
Cypress *(Cupressus sempervirens)*	-
Elemi *(Canarium luzonicum)*	0–0.1%
Eucalyptus citriodora	67–86%
Eucalyptus dives	tr
Eucalyptus globulus (not rectified)	tr
Eucalyptus radiata	0–0.4%

Essential Oils	Total Range Aldehydes
Eucalyptus staigeriana	6.7–19.3%
Fennel *(Foeniculum vulgare)*	0–0.4%.
Fir, Douglas *(Pseudotsuga menziesii)*	-
Frankincense *(Boswellia carterii)*	0–1.5%
Galbanum *(Ferula galbaniflua)*	-
Geranium *(Pelargonium x asperum)*	0–0.6%
Ginger *(Zinziber officinale)*	0–5%
Grapefruit *(Citrus paradisi)*	0–0.4%
Helichrysum *(Helichrysum italicum)*	-
Inula graveolens	-
Hyssop *(Hyssopus officinalis)*	-
Jasmine, Absolute *(Jasminum grandiflorum)*	-
Juniper *(Juniperus communis)*	-

Essential Oils	Total Range Aldehydes
Kunzea ambigua	tr–0.3%
Lavandin *(Lavandula hybrida)*	-
Lavender *(Lavandula angustifolia)*	0–0.9%
Lemon *(Citrus limon)*	0.5–4.5%
Lemongrass *(Cymbopogon citratus)*	56.7–80%
Lime *(Citrus limetta)*	tr–2.5%
Litsea cubeba	64.4–73.6%
Manadrin *(Citrus reticulata)* peel	tr–0.5%
Mandarin Petitgrain *(Citrus reticulata)* leaves	-
Marjoram *(Origanum hortensis)*	0–0.12%
Melissa *(Melissa officinalis)*	26.7–54%
MQV *(Melaleuca quinquenervia viridiflora)* CT viridiflorol	-
Myrrh *(Commiphora myrrha)*	-

Essential Oils	Total Range Aldehydes
Myrtle, Green *(Myrtus communis)*	-
Neroli *(Citrus aurantium)* blossom	-
Nutmeg *(Myristica fragrans)*	-
Orange, Bitter *(Citrus aurantium)* peel	tr–1.2%
Orange, Sweet *(Citrus sinensis)*	0.5–3%
Oregano *(Origanum vulgaris)*	-
Palmarosa *(Cymbopogon martini)*	0.5–1.9%
Patchouli *(Pogostemon cablin)*	-
Pepper Black *(Piper nigrum)*	-
Peppermint *(Mentha piperita)*	-
Petitgrain *(Citrus aurantium)* leaves	-
Pine *(Pinus sylvestris)*	-
Plai *(Zingiber cassumunar)*	-
Ravensare *(Ravensara aromatica)*	-

Essential Oils	Total Range Aldehydes
Ravintsara/Ho Leaf CT Cineole *(Cinnamomum camphora)*	-
Rose *(Rosa damascena)*	0.8–2.5%
Rosemary, CT Cineole *(Rosmarinus officinalis)*	-
Rosemary, CT Verbenone *(Rosmarinus officinalis)*	-
Sage *(Salvia officinalis)*	-
Saint John's Wort *(Hypericum perforatum)*	
Sandalwood, Australian *(Santalum spicatum)*	-
Spearmint *(Mentha viridis)*	-
Spike Lavender *(Lavandula latifolia)*	-
Spikenard *(Nardostachys jatamansi)*	-
Spruce, Black *(Picea mariana)*	-

Essential Oils	Total Range Aldehydes
Tarragon *(Artemisia dracunculus)*	0–0.4%
Tea Tree *(Melaleuca alternifolia)*	tr
Thyme, CT Geraniol *(Thymus vulgaris)*	0–0.8%
Thyme, CT Linalool *(Thymus vulgaris)*	-
Thyme, CT Thymol *(Thymus vulgaris)*	0–tr
Thymus satureioides	-
Vetiver *(Vetiveria zizanoides)*	-
Vitex *(Vitex agnus castus)*	-
Yarrow CT Chamazulene *(Achillea millefolium)*	tr–1%
Ylang Ylang, Complete *(Cananga odorata)*	-

Ketones

Essential Oils	Total Range Ketones
Anise Seed *(Pimpinella anisum)*	tr
Balsam Poplar *(Populus balsamifera)*	-
Basil, Sweet, CT Linalool *(Ocimum basilicum)*	0.1–1%
Bay Laurel *(Laurus nobilis)*	-
Bergamot *(Citrus aurantium bergamia)*	-
Blue Tansy *(Tanacetum annuum)*	3.1–14%
Cardamon *(Elettaria cardamomum)*	-
Carrot Seed *(Daucus carota)*	0–0.02%
Cedarwood, Atlas *(Cedrus atlantica)*	8–28.5%
Cedarwood, Virginia *(Juniperus virginiana)*	0–0.2%
Chamomile, Cape *(Eriocephalus punctulatus)*	-
Chamomile, German *(Matricaria recutita)*	tr

Essential Oils	Total Range Ketones
Chamomile, Roman *(Chamaemelum nobile)*	0–2.4%
Cinnamon, bark *(Cinnamomum zeylanicum)*	tr–1.4%
Cistus *(Cistus ladaniferus)*	5–8.7%
Clary Sage *(Salvia sclarea)*	-
Clove Bud *(Eugenia caryophyllata)*	-
Copaiba *(Copaifera officianalis)*	-
Coriander *(Coriandrum sativum)*	2–7.7%
Cypress *(Cupressus sempervirens)*	-
Elemi *(Canarium luzonicum)*	0–0.2%
Eucalyptus citriodora	-
Eucalyptus dives	35.5–54%
Eucalyptus globulus (not rectified)	tr–1%
Eucalyptus radiata	0.4–4.7%

Essential Oils	Total Range Ketones
Eucalyptus staigeriana	-
Fennel *(Foeniculum vulgare)*	0.2–8.0%
Fir, Douglas *(Pseudotsuga menziesii)*	tr–1.4%
Frankincense *(Boswellia carterii)*	0.3–0.4%
Galbanum *(Ferula galbaniflua)*	-
Geranium *(Pelargonium x asperum)*	3–7.8%
Ginger *(Zinziber officinale)*	0–1.4%
Grapefruit *(Citrus paradisi)*	tr–0.8%
Helichrysum *(Helichrysum italicum)*	6–8%
Inula graveolens	0–0.2%
Hyssop *(Hyssopus officinalis)*	60–72%
Jasmine Absolute *(Jasminum grandiflorum)*	0–5%
Juniper *(Juniperus communis)*	-
Kunzea ambigua	-

Essential Oils	Total Range Ketones
Lavandin *(Lavandula hybrida)*	4.5–12%
Lavender *(Lavandula angustifolia)*	1–2.5%
Lemon *(Citrus limon)*	tr
Lemongrass *(Cymbopogon citratus)*	tr–3%
Lime *(Citrus limetta)*	-
Litsea cubeba	0.5–4.4%
Manadrin *(Citrus reticulata)* peel	-
Mandarin Petitgrain *(Citrus reticulata)* leaves	-
Marjoram *(Origanum hortensis)*	2–3%
Melissa *(Melissa officinalis)*	0–2.5%
MQV *(Melaleuca quinquenervia viridiflora)* CT viridiflorol	-
Myrrh *(Commiphora myrrha)*	0–1.1%
Myrtle, Green *(Myrtus communis)*	-

Essential Oils	Total Range Ketones
Neroli *(Citrus aurantium)* blossom	-
Nutmeg *(Myristica fragrans)*	-
Orange, Bitter *(Citrus aurantium)* peel	-
Orange, Sweet *(Citrus sinensis)*	1–0.5%
Oregano *(Origanum vulgaris)*	-
Palmarosa *(Cymbopogon martini)*	-
Patchouli *(Pogostemon cablin)*	0–5%
Pepper, Black *(Piper nigrum)*	-
Peppermint *(Mentha piperita)*	25–34%
Petitgrain *(Citrus aurantium)* leaves	-
Pine *(Pinus sylvestris)*	0–2.1%
Plai *(Zingiber cassumunar)*	-
Ravensare *(Ravensara aromatica)*	-

Essential Oils	Total Range Ketones
Ravintsara/Ho Leaf CT Cineole *(Cinnamomum camphora)*	-
Rose *(Rosa damascena)*	-
Rosemary, CT Cineole *(Rosmarinus officinalis)*	8–14.9%
Rosemary, CT Verbenone *(Rosmarinus officinalis)*	7.5–27%
Sage *(Salvia officinalis)*	16–50.2%
Saint John's Wort *(Hypericum perforatum)*	0–3.2%
Sandalwood, Australian *(Santalum spicatum)*	-
Spearmint *(Mentha viridis)*	66–72%
Spike Lavender *(Lavandula latifolia)*	10.8–23.2%
Spikenard *(Nardostachys jatamansi)*	6–9%
Spruce, Black *(Picea mariana)*	0.2–4.9%

Essential Oils	Total Range Ketones
Tarragon *(Artemisia dracunculus)*	-
Tea Tree *(Melaleuca alternifolia)*	-
Thyme, CT Geraniol *(Thymus vulgaris)*	-
Thyme, CT Linalool *(Thymus vulgaris)*	tr–0.5%
Thyme, CT Thymol *(Thymus vulgaris)*	0–1.7%
Thymus satureioides	0–1%
Vetiver *(Vetiveria zizanoides)*	2–6.4%
Vitex *(Vitex agnus castus)*	-
Yarrow CT Chamazulene *(Achillea millefolium)*	0.5–7.5%
Ylang Ylang, Complete *(Cananga odorata)*	-

Lactones

Essential Oils	Total Range Lactones
Anise Seed *(Pimpinella anisum)*	-
Balsam Poplar *(Populus balsamifera)*	-
Basil, Sweet, CT Linalool *(Ocimum basilicum)*	-
Bay Laurel *(Laurus nobilis)*	-
Bergamot *(Citrus aurantium bergamia)*	0.3–4%
Blue Tansy *(Tanacetum annuum)*	-
Cardamon *(Elettaria cardamomum)*	-
Carrot Seed *(Daucus carota)*	-
Cedarwood, Atlas *(Cedrus atlantica)*	-
Cedarwood, Virginia *(Juniperus virginiana)*	-
Chamomile, Cape *(Eriocephalus punctulatus)*	tr–1.7%
Chamomile, German *(Matricaria recutita)*	-

Essential Oils	Total Range Lactones
Chamomile, Roman *(Chamaemelum nobile)*	-
Cinnamon, bark *(Cinnamomum zeylanicum)*	-
Cistus *(Cistus ladaniferus)*	-
Clary Sage *(Salvia sclarea)*	-
Clove Bud *(Eugenia caryophyllata)*	-
Copaiba *(Copaifera officianalis)*	-
Coriander *(Coriandrum sativum)*	-
Cypress *(Cupressus sempervirens)*	-
Elemi *(Canarium luzonicum)*	-
Eucalyptus citriodora	-
Eucalyptus dives	-
Eucalyptus globulus (not rectified)	-
Eucalyptus radiata	-

Essential Oils	Total Range Lactones
Eucalyptus staigeriana	-
Fennel *(Foeniculum vulgare)*	-
Fir, Douglas *(Pseudotsuga menziesii)*	-
Frankincense *(Boswellia carterii)*	-
Galbanum *(Ferula galbaniflua)*	0–1.4%
Geranium *(Pelargonium x asperum)*	-
Ginger *(Zinziber officinale)*	-
Grapefruit *(Citrus paradisi)*	tr–0.4%
Helichrysum *(Helichrysum italicum)*	-
Inula graveolens	4%
Hyssop *(Hyssopus officinalis)*	-
Jasmine Absolute *(Jasminum grandiflorum)*	-
Juniper *(Juniperus communis)*	-
Kunzea ambigua	-
Lavandin *(Lavandula hybrida)*	-
Lavender *(Lavandula angustifolia)*	tr
Lemon *(Citrus limon)*	tr–3%
Lemongrass *(Cymbopogon citratus)*	-
Lime *(Citrus limetta)*	2–8%
Litsea cubeba	-
Manadrin *(Citrus reticulata)* peel	tr
Mandarin Petitgrain *(Citrus reticulata)* leaves	-
Marjoram *(Origanum hortensis)*	-
Melissa *(Melissa officinalis)*	-
MQV *(Melaleuca quinquenervia viridiflora)* CT viridiflorol	-
Myrrh *(Commiphora myrrha)*	59%
Myrtle, Green *(Myrtus communis)*	-

Essential Oils	Total Range Lactones
Neroli *(Citrus aurantium)* blossom	-
Nutmeg *(Myristica fragrans)*	-
Orange, Bitter *(Citrus aurantium)* peel	-
Orange, Sweet *(Citrus sinensis)*	tr–0.2%
Oregano *(Origanum vulgaris)*	-
Palmarosa *(Cymbopogon martini)*	-
Patchouli *(Pogostemon cablin)*	-
Pepper, Black *(Piper nigrum)*	-
Peppermint *(Mentha piperita)*	1–6%
Petitgrain *(Citrus aurantium)* leaves	-
Pine *(Pinus sylvestris)*	-
Plai *(Zingiber cassumunar)*	-
Ravensare *(Ravensara aromatica)*	-

Essential Oils	Total Range Lactones
Ravintsara/Ho Leaf CT Cineole *(Cinnamomum camphora)*	-
Rose *(Rosa damascena)*	-
Rosemary, CT Cineole *(Rosmarinus officinalis)*	-
Rosemary, CT Verbenone *(Rosmarinus officinalis)*	-
Sage *(Salvia officinalis)*	-
Saint John's Wort *(Hypericum perforatum)*	-
Sandalwood, Australian *(Santalum spicatum)*	3–5.3%
Spearmint *(Mentha viridis)*	-
Spike Lavender *(Lavandula latifolia)*	-
Spikenard *(Nardostachys jatamansi)*	-
Spruce, Black *(Picea mariana)*	-

Essential Oils	Total Range Lactones
Tarragon *(Artemisia dracunculus)*	-
Tea Tree *(Melaleuca alternifolia)*	-
Thyme, CT Geraniol *(Thymus vulgaris)*	-
Thyme, CT Linalool *(Thymus vulgaris)*	-
Thyme, CT Thymol *(Thymus vulgaris)*	-
Thymus satureioides	-
Vetiver *(Vetiveria zizanoides)*	0–0.2%
Vitex *(Vitex agnus castus)*	-
Yarrow CT Chamazulene *(Achillea millefolium)*	-
Ylang Ylang, Complete *(Cananga odorata)*	-

Oxides

Essential Oils	Total Range Oxides
Anise Seed *(Pimpinella anisum)*	-
Balsam Poplar *(Populus balsamifera)*	-
Basil, Sweet, CT Linalool *(Ocimum basilicum)*	4.9–7.2%
Bay Laurel *(Laurus nobilis)*	38.1–61%
Bergamot *(Citrus aurantium bergamia)*	-
Blue Tansy *(Tanacetum annuum)*	0.4–1.7%
Cardamon *(Elettaria cardamomum)*	26.5–44.6%
Carrot Seed *(Daucus carota)*	0.3–2.8%
Cedarwood, Atlas *(Cedrus atlantica)*	0–1.6%
Cedarwood, Virginia *(Juniperus virginiana)*	-
Chamomile, Cape *(Eriocephalus punctulatus)*	3–7.1%
Chamomile, German *(Matricaria recutita)*	3–53%

Essential Oils	Total Range Oxides
Chamomile, Roman *(Chamaemelum nobile)*	-
Cinnamon, bark *(Cinnamomum zeylanicum)*	0.4–4.6%
Cistus *(Cistus ladaniferus)*	-
Clary Sage *(Salvia sclarea)*	1–1.6%
Clove Bud *(Eugenia caryophyllata)*	0–1%
Copaiba *(Copaifera officianalis)*	-
Coriander *(Coriandrum sativum)*	-
Cypress *(Cupressus sempervirens)*	0–0.4%
Elemi *(Canarium luzonicum)*	0–2.5%
Eucalyptus citriodora	tr
Eucalyptus dives	1.7–2.5%
Eucalyptus globulus (not rectified)	65.4–72%
Eucalyptus radiata	60.4–69%
Eucalyptus staigeriana	4–6.9%

Essential Oils	Total Range Oxides
Fennel *(Foeniculum vulgare)*	-
Fir, Douglas *(Pseudotsuga menziesii)*	-
Frankincense *(Boswellia carterii)*	tr–0.3%
Galbanum *(Ferula galbaniflua)*	-
Geranium *(Pelargonium x asperum)*	0.4–4.4%
Ginger *(Zinziber officinale)*	0.2–3.3%
Grapefruit *(Citrus paradisi)*	0–0.5%
Helichrysum *(Helichrysum italicum)*	tr–3%
Inula graveolens	0–5.03%
Hyssop *(Hyssopus officinalis)*	-
Jasmine Absolute *(Jasminum grandiflorum)*	-
Juniper *(Juniperus communis)*	-
Kunzea ambigua	13–14.8%

Essential Oils	Total Range Oxides
Lavandin *(Lavandula hybrida)*	3–10.4%
Lavender *(Lavandula angustifolia)*	0–1.4%
Lemon *(Citrus limon)*	0–0.4%
Lemongrass *(Cymbopogon citratus)*	0–6%
Lime *(Citrus limetta)*	tr–2%
Litsea cubeba	-
Manadrin *(Citrus reticulata)* peel	-
Mandarin Petitgrain *(Citrus reticulata)* leaves	-
Marjoram *(Origanum hortensis)*	0–2.1%
Melissa *(Melissa officinalis)*	0.8–10%
MQV *(Melaleuca quinquenervia viridiflora)* CT viridiflorol	30–35%
Myrrh *(Commiphora myrrha)*	-
Myrtle, Green *(Myrtus communis)*	19–37.5%

Essential Oils	Total Range Oxides
Neroli *(Citrus aurantium)* blossom	-
Nutmeg *(Myristica fragrans)*	0–tr
Orange, Bitter *(Citrus aurantium)* peel	-
Orange, Sweet *(Citrus sinensis)*	-
Oregano *(Origanum vulgaris)*	0–1.4%
Palmarosa *(Cymbopogon martini)*	0.1–1.8%
Patchouli *(Pogostemon cablin)*	0–2.7%
Pepper, Back *(Piper nigrum)*	0–0.7%
Peppermint *(Mentha piperita)*	2–6%
Petitgrain *(Citrus aurantium)* leaves	-
Pine *(Pinus sylvestris)*	0–7%
Plai *(Zingiber cassumunar)*	0–0.2%
Ravensare *(Ravensara aromatica)*	1.8–3.3%

Essential Oils	Total Range Oxides
Ravintsara/Ho Leaf CT Cineole *(Cinnamomum camphora)*	56–63%
Rose *(Rosa damascena)*	0.1–0.4%
Rosemary, CT Cineole *(Rosmarinus officinalis)*	39.0–57.7%
Rosemary, verbenone type *(Rosmarinus officinalis)*	0.3–9%
Sage *(Salvia officinalis)*	2–21%
Saint John's Wort *(Hypericum perforatum)*	0–0.9%
Sandalwood, Australian *(Santalum spicatum)*	-
Spearmint *(Mentha viridis)*	0.7–1.5%
Spike Lavender *(Lavandula latifolia)*	19–34.9%
Spikenard *(Nardostachys jatamansi)*	-
Spruce, Black *(Picea mariana)*	0–0.5%

Essential Oils	Total Range Oxides
Tarragon *(Artemisia dracunculus)*	-
Tea Tree *(Melaleuca alternifolia)*	3–4%
Thyme, CT Geraniol *(Thymus vulgaris)*	0–0.2%
Thyme, CT Linalool *(Thymus vulgaris)*	0.4–4%
Thyme, CT Thymol *(Thymus vulgaris)*	-
Thymus satureioides	0–2%
Vetiver *(Vetiveria zizanoides)*	16–39%
Vitex *(Vitex agnus castus)*	-
Yarrow CT Chamazulene *(Achillea millefolium)*	0–4.4%
Ylang Ylang Ylang, Complete *(Cananga odorata)*	0–0.3%

Phenylpropanoids (Hot)

Essential Oils	Total Range Hot Phenylpropanoids
Anise Seed *(Pimpinella anisum)*	-
Balsam Poplar *(Populus balsamifera)*	-
Basil, Sweet, CT Linalool *(Ocimum basilicum)*	tr–1.6%
Bay Laurel *(Laurus nobilis)*	0.7–2.9%
Bergamot *(Citrus aurantium bergamia)*	-
Blue Tansy *(Tanacetum annuum)*	tr
Cardamon *(Elettaria cardamomum)*	-
Carrot Seed *(Daucus carota)*	-
Cedarwood, Atlas *(Cedrus atlantica)*	-
Cedarwood, Virginia *(Juniperus virginiana)*	-
Chamomile, Cape *(Eriocephalus punctulatus)*	-
Chamomile, German *(Matricaria recutita)*	-
Chamomile, Roman *(Anthemis nobilis)*	-
Cinnamon, bark *(Cinnamomum zeylanicum)*	63–90%
Cistus *(Cistus ladaniferus)*	tr–2.3%
Clary Sage *(Salvia sclarea)*	-
Clove Bud *(Eugenia caryophyllata)*	73.6–97.4%
Copaiba *(Copaifera officianalis)*	-
Coriander *(Coriandrum sativum)*	-
Cypress *(Cupressus sempervirens)*	-
Elemi *(Canarium luzonicum)*	-
Eucalyptus citriodora	-
Eucalyptus dives	-
Eucalyptus globulus (not rectified)	-
Eucalyptus radiata	-
Eucalyptus staigeriana	-

Essential Oils	Total Range Hot Phenylpro-anoids
Fennel *(Foeniculum vulgare)*	-
Fir, Douglas *(Pseudotsuga menziesii)*	-
Frankincense *(Boswellia carterii)*	-
Galbanum *(Ferula galbaniflua)*	-
Geranium *(Pelargonium x asperum)*	-
Ginger *(Zinziber officinale)*	-
Grapefruit *(Citrus paradisi)*	-
Helichrysum *(Helichrysum italicum)*	-
Inula graveolens	-
Hyssop *(Hyssopus officinalis)*	-
Jasmine Absolute *(Jasminum grandiflorum)*	1.1–3%
Juniper *(Juniperus communis)*	-
Kunzea ambigua	-

Essential Oils	Total Range Hot Phenylpro-anoids
Lavandin *(Lavandula hybrida)*	-
Lavender *(Lavandula angustifolia)*	-
Lemon *(Citrus limon)*	-
Lemongrass *(Cymbopogon citratus)*	-
Lime *(Citrus limetta)*	-
Litsea cubeba	-
Manadrin *(Citrus reticulata)* peel	-
Mandarin Petitgrain *(Citrus reticulata)* leaves	-
Marjoram *(Origanum hortensis)*	-
Melissa *(Melissa officinalis)*	-
MQV *(Melaleuca quinquenervia viridiflora)* CT viridiflorol	-
Myrrh *(Commiphora myrrha)*	-
Myrtle, Green *(Myrtus communis)*	-

Essential Oils	Total Range Hot Phenylpro-anoids
Neroli *(Citrus aurantium)* blossom	-
Nutmeg *(Myristica fragrans)*	-
Orange, Bitter *(Citrus aurantium)* peel	-
Orange, Sweet *(Citrus sinensis)*	-
Oregano *(Origanum vulgaris)*	-
Palmarosa *(Cymbopogon martini)*	-
Patchouli *(Pogostemon cablin)*	-
Pepper, Black *(Piper nigrum)*	-
Peppermint *(Mentha piperita)*	-
Petitgrain *(Citrus aurantium)* leaves	-
Pine *(Pinus sylvestris)*	-
Plai *(Zingiber cassumunar)*	-
Ravensare *(Ravensara aromatica)*	0–0.8%

Essential Oils	Total Range Hot Phenylpro-anoids
Ravintsara/Ho Leaf CT Cineole *(Cinnamomum camphora)*	-
Rose *(Rosa damascena)*	0.5–1.2%
Rosemary, CT Cineole *(Rosmarinus officinalis)*	-
Rosemary, CT Verbenone *(Rosmarinus officinalis)*	-
Sage *(Salvia officinalis)*	-
Saint John's Wort *(Hypericum perforatum)*	-
Sandalwood, Australian *(Santalum spicatum)*	-
Spearmint *(Mentha viridis)*	-
Spike Lavender *(Lavandula latifolia)*	-
Spikenard *(Nardostachys jatamansi)*	-
Spruce, Black *(Picea mariana)*	-

Essential Oils	Total Range Hot Phenylpro-anoids
Tarragon *(Artemisia dracunculus)*	-
Tea Tree *(Melaleuca alternifolia)*	-
Thyme, CT Geraniol *(Thymus vulgaris)*	-
Thyme, CT Linalool *(Thymus vulgaris)*	-
Thyme, CT Thymol *(Thymus vulgaris)*	-
Thymus satureioides	-
Vetiver *(Vetiveria zizanoides)*	-
Vitex *(Vitex agnus cas-tus)*	-
Yarrow CT Chamazu-lene *(Achillea millefo-lium)*	tr
Ylang Ylang, Complete *(Cananga odorata)*	0–0.7%

Ethers

Essential Oils	Total Range Ethers
Anise Seed *(Pimpinella anisum)*	75–96.5%
Balsam Poplar *(Populus balsamifera)*	-
Basil, Sweet, CT Linalool *(Ocimum basilicum)*	0.05–9%
Bay Laurel *(Laurus nobilis)*	1.4–3.8%
Bergamot *(Citrus aurantium bergamia)*	-
Blue Tansy *(Tanacetum annuum)*	-
Cardamon *(Elettaria cardamomum)*	-
Carrot Seed *(Daucus carota)*	-
Cedarwood, Atlas *(Cedrus atlantica)*	-
Cedarwood, Virginia *(Juniperus virginiana)*	-
Chamomile, Cape *(Eriocephalus punctulatus)*	-
Chamomile, German *(Matricaria recutita)*	-
Chamomile, Roman (Anthemis nobilis)	-
Cinnamon, bark *(Cinnamomum zeylanicum)*	0–0.4%
Cistus *(Cistus ladaniferus)*	-
Clary Sage *(Salvia sclarea)*	-
Clove Bud *(Eugenia caryophyllata)*	0–.2%
Copaiba *(Copaifera officianalis)*	-
Coriander *(Coriandrum sativum)*	-
Cypress *(Cupressus sempervirens)*	-
Elemi *(Canarium luzonicum)*	2–10.6%
Eucalyptus citriodora	-
Eucalyptus dives	-
Eucalyptus globulus (not rectified)	-
Eucalyptus radiata	-
Eucalyptus staigeriana	-

Essential Oils	Total Range Ethers
Fennel *(Foeniculum vulgare)*	58.2–97.3%
Fir, Douglas *(Pseudotsuga menziesii)*	-
Frankincense *(Boswellia carterii)*	0–4%
Galbanum *(Ferula galbaniflua)*	-
Geranium *(Pelargonium x asperum)*	-
Ginger *(Zinziber officinale)*	-
Grapefruit *(Citrus paradisi)*	-
Helichrysum *(Helichrysum italicum)*	-
Inula graveolens	0–0.3%
Hyssop *(Hyssopus officinalis)*	0–4%
Jasmine Absolute *(Jasminum grandiflorum)*	-
Juniper *(Juniperus communis)*	-
Kunzea ambigua	-

Essential Oils	Total Range Ethers
Lavandin *(Lavandula hybrida)*	-
Lavender *(Lavandula angustifolia)*	-
Lemon *(Citrus limon)*	-
Lemongrass *(Cymbopogon citratus)*	0–.2%
Lime *(Citrus limetta)*	-
Litsea cubeba	-
Manadrin *(Citrus reticulata)* peel	-
Mandarin Petitgrain *(Citrus reticulata)* leaves	-
Marjoram, Sweet *(Origanum hortensis)*	-
Melissa *(Melissa officinalis)*	tr
MQV *(Melaleuca quinquenervia viridiflora)* CT viridiflorol	-
Myrrh *(Commiphora myrrha)*	-
Myrtle, Green *(Myrtus communis)*	0.3–2.6%

Essential Oils	Total Range Ethers
Neroli *(Citrus aurantium)* blossom	-
Nutmeg *(Myristica fragrans)*	9–22%
Orange, Bitter *(Citrus aurantium)* peel	-
Orange, Sweet *(Citrus sinensis)*	-
Oregano *(Origanum vulgaris)*	-
Palmarosa *(Cymbopogon martini)*	-
Patchouli *(Pogostemon cablin)*	-
Pepper, Black *(Piper nigrum)*	-
Peppermint *(Mentha piperita)*	-
Peru Balsam *(Myroxylon pereira)*	
Petitgrain *(Citrus aurantium)* leaves	-
Pine *(Pinus sylvestris)*	-
Plai *(Zingiber cassumunar)*	-
Ravensare *(Ravensara aromatica)*	2.4–11.9%
Ravintsara/Ho Leaf CT Cineole *(Cinnamomum camphora)*	0–0.2%
Rose *(Rosa damascena)*	0.5–3.3%
Rosemary, CT Cineole *(Rosmarinus officinalis)*	-
Rosemary, verbenone type *(Rosmarinus officinalis)*	-
Sage *(Salvia officinalis)*	-
Saint John's Wort *(Hypericum perforatum)*	-
Sandalwood, Australian *(Santalum spicatum)*	-
Spearmint *(Mentha viridis)*	-
Spike Lavender *(Lavandula latifolia)*	-
Spikenard *(Nardostachys jatamansi)*	-

Essential Oils	Total Range Ethers
Spruce, Black *(Picea mariana)*	-
Tarragon *(Artemisia dracunculus)*	73.4–88.8%
Tea Tree *(Melaleuca alternifolia)*	-
Thyme, CT Geraniol *(Thymus vulgaris)*	-
Thyme, CT Linalool *(Thymus vulgaris)*	-
Thyme, CT Thymol *(Thymus vulgaris)*	0–1.4%
Thymus satureioides	1–4%
Vetiver *(Vetiveria zizanoides)*	0–0.6%
Vitex *(Vitex agnus castus)*	-
Yarrow CT Chamazulene *(Achillea millefolium)*	-
Ylang Ylang, Complete *(Cananga odorata)*	0.2 - 3.3%

ABOUT THE AUTHOR

Jimm Harrison founded the Phytotherapy Institute in 1995 to advance education in essential oils and plant therapy. He is a published author and has conducted certification programs and lectures in essential oil therapy for natural health, massage, spa therapy, esthetics and medical institutions across the U.S. and Internationally, including Harvard University, Dana-Farber Cancer Institute and Harvard Medical School.

As a leading authority in natural skin health and essential oils he has helped to pioneer nutritional holistic beauty programs. Jimm is an expert formulator and has customized essential oil systems for spa and resorts, formulated essential oil products and has developed professional and retail organic, nutritional skin care.

He is the author of *Aromatherapy: Therapeutic Use of Essential Oils for Esthetics* and developed and teaches the Essential Oil and Aromatherapy Certificate Program for Bastyr University CCCE.

Learn more: **www.jimmharrison.com**

Frankincense Tree in Oman

PHOTO CREDIT © SHUTTERSTOCK

CPSIA information can be obtained
at www.ICGtesting.com
Printed in the USA
LVHW071721271119
638725LV00004B/185/P

* 9 7 8 0 9 8 6 3 5 3 9 3 2 *